AF757397

Fleeing the City

Studies in the Culture and Politics of Antiurbanism

Edited By
Michael J. Thompson

palgrave
macmillan

FLEEING THE CITY
Copyright © Michael J. Thompson, 2009.

All rights reserved.

First published in 2009 by
PALGRAVE MACMILLAN®
in the United States—a division of St. Martin's Press LLC,
175 Fifth Avenue, New York, NY 10010.

Where this book is distributed in the UK, Europe and the rest of the world, this is by Palgrave Macmillan, a division of Macmillan Publishers Limited, registered in England, company number 785998, of Houndmills, Basingstoke, Hampshire RG21 6XS.

Palgrave Macmillan is the global academic imprint of the above companies and has companies and representatives throughout the world.

Palgrave® and Macmillan® are registered trademarks in the United States, the United Kingdom, Europe and other countries.

ISBN: 978–0–230–61059–0

Library of Congress Cataloging-in-Publication Data

Fleeing the city : studies in the culture and politics of antiurbanism / edited by Michael J. Thompson.
p. cm.
Includes bibliographical references and index.
ISBN 0–230–61059–5
1. Urbanization. 2. Urban policy. 3. Country life. I. Thompson, Michael, 1973–

HT361.F53 2009
307.76—dc22 2009002673

A catalogue record of the book is available from the British Library.

Design by Newgen Imaging Systems (P) Ltd., Chennai, India.

First edition: September 2009

10 9 8 7 6 5 4 3 2 1

Printed in the United States of America.

CONTENTS

Part 3 Antiurbanism in Society and Politics

FIGURES

TABLES

Introduction

Antiurbanism has received scant attention in the social scientific and historical literature. As a theme in political, cultural, and social history as well as a political and sociological phenomenon, antiurbanism has been sadly understudied and undertheorized.[1] Generally it is seen as a product of the fear of cities and their anomic consequences on moral life. Conservative reaction to modernity also included an antiurban impulse since it was in cities that the impact of technological progress could be seen in the ways that workers and families disintegrated and the traditional morality of the past was called into question both in theory and in practice. In these instances, antiurbanism was seen as a means toward the regeneration of traditional life and values. All of this makes sense, but the deeper valences of this problem have not been adequately addressed. The urbanization of consciousness has given rise, historically as well as today, to an *antiurbanized* consciousness; a sense that cities and urban life more broadly are to be uniformly feared. In this respect, although antiurbanism has largely been studied as a historical reality—one confined to the reaction to the rise of modernity and industrialism—it has by no means subsided in postindustrial societies. In this respect, understanding the ways that antiurbanism has manifested itself in thought and practice is crucial, but it is also important to go a step further: to probe the deeper dimensions of antiurban sentiment in consciousness, in institutional logics, and broader patterns of cultural production. This is the aim of the present volume.

As an idea, as a tradition, antiurbanism has changed and evolved over time. In many ways, the essence of antiurbanism is the distinction between what is "natural" as opposed to that which is, in some way, "artificial." The distinction is a very dynamic one, but it essentially covers the ideological field within which the antiurban impulse operates. This, however, was an ideology that evolved over time—the complexities of city life as opposed to rural life are not historically persistent. The idea that the city was unnatural, however, was not something inherent to the process of cities themselves. In the ancient Greek world, Aristotle was clear that the *polis* was the highest and most perfect form of human organization and association. The *polis* was, in the ancient Greek mind, necessary for human development—for the betterment of the self and the production of the best faculties of humankind. Aristotle's notion of the *polis* as the most perfect form of human association was mirrored by his notion of it as "inherently natural": man could only evolve as truly human—as opposed to Barbarian—within the dense ties of complex association that the polis could provide. Cities became known not only as centers of political and economic power, but also as places of cultural production and the realm of human fulfillment and happiness.

The contrary to this in the ancient world was the pastoral—the escape from the corruption of places like Athens and Rome into a place of simplicity and peace. This manifested itself mainly in literary genres and was not a robust political tradition or cultural mindset.[2] The real roots of antiurban thinking emerge as a reaction not to the "modern" city, as many have assumed, but rather to the feudal city during the seventeenth century.[3] Antiurbanism did not begin as a regressive force in the modern sense of the term. In England during the seventeenth and eighteenth centuries, the division between the countryside and the city took on deep political valences. Radical Protestantism tied individual labor to property and to virtue; its political ideology was a function of both the puritanism of English country life as well as the early manufacturing that was emerging first outside of the cities and in the surrounding countryside.[4] This was a full expression of antiurbanism, and it was deeply wedded to religious, political, and economic themes that associated the city with idleness, corruption, and debauchery. The English revolutions of the seventeenth century therefore understandably originated in nonurban regions, and the reaction against the city was tied to a liberal reaction against privilege and feudal orders of rank just as much as it was tied to Protestant morality. It was a political and cultural response to the decadence of urban life, of excessive wealth, money worship, and other vices. This form of antiurbanism was a religiously inspired political movement with deep cultural consequences. Indeed, it would be carried over into the New World and also plant the seeds of an American form of antiurbanism in the ideas of Thomas Jefferson, Henry David Thoreau, Henry Adams, Robert Park, and Frank Lloyd Wright.[5]

In Europe, Romanticism was defined by its flight from urban life. The city just as the manufacturing town was a place of falsity and alienation. It was a place of conformity, of rigid quantification and commodification. As a response to the emerging urban industrialism, English Romanticism evoked not only nature as an aesthetic ideal, but as an opposition to the corruption of modern life itself. The massive transformation of time and space that accompanied the industrial revolution and the massive urbanization that was its counterpart made this inevitable. The idea of the city was treated with disdain for the simple reason that it was degrading human life. In his *Emile*, Rousseau shows how the city is coupled with the corruption of man:

> [W]hen the critical age approaches, furnish young people with sights which restrain them and not with sights which arouse them. Put their nascent imaginations off the track with objects which, far from inflaming, repress the activity of their senses. Remove them from big cities where the adornment and the immodesty of women hasten and anticipate nature's lessons, where everything presents to their eyes pleasures they ought to know only when they are able to choose among them. Bring them back to their first abodes where rustic simplicity lets the passions of their age develop less rapidly. Or if their taste for the arts still attaches them to the city, keep them from a dangerous idleness by means of this very taste.[6]

For, Rousseau, the antiurban impulse becomes a reaction not simply to transformations in economic life, but to the very way that the ways social relations were transforming within urban centers. The reaction against the city was largely characterized by the rejection of complex forms of social relations and their cultural implications. It grew into a reaction against modern forms of life and the forms of consciousness and culture to which urbanization gave rise. Cities became viewed as collections of rootless individuals characterized by competition, the impersonal, and conflict. It was a place not for man's development, but for his corruption, his debasement.

Since the emergence of modernity was inherently coupled with urbanism, the reaction against modernity would increasingly become associated with the antiurban impulse. The expansion of industrial capitalism brought with it urbanization, population density, and immigration, challenging traditional conceptions of community, ethnicity, sexual identity, and the family. Tied with the culture of cities was a corresponding breakdown of traditional forms of life. Whereas nonurban areas were characterized by familial and religious bonds and primary social relations, urban areas saw the breakdown of these bonds and created a culture that was decidedly different, modern in orientation. The antiurban impulse became more intense with the deepening of the culture of urban life and the ways that it was alienating individuals from "authenticity," however variously that was defined. Different forms of association therefore led to different forms of consciousness and, in the end, to different ideologies concerning culture and politics. In Europe, antiurbanism became a hallmark of fascist movements that seized upon the anarchic, rootless nature of urban life to promote a movement away from modern forms of consciousness—bourgeois individualism, democratic forms of association, and so on.[7] Conservatives and Romantics of all stripes began to respond to the corrupting, dehumanizing effects of urban existence and the threats it posed to traditional forms of communal morality and authority. And all of this makes sense since antiurbanism was a direct product of the erosion of cohesive moral fabric that had always held the human social order together, legitimizing traditional orders within the political and familial spheres as well as the various identities and institutions that held those orders together.

Since cities tend to be the locus of cultural change, of dynamism and social evolution, they also tend to be perceived as threats to the moral fabric of tradition and custom; they provide space for social and cultural innovation and, as a result, fray the fabric of traditional, communal forms of life and institutions. For this reason, even outside of the developmental history of the West, antiurbanism has also been a theme of contemporary movements in the Middle East and elsewhere in the non-Western world, and it is not difficult to see why. Globalization is now creating similar social, political, economic, and cultural pressures and forces due to increasing urbanization in the developing world, and it is no coincidence that we see antiurbanism emerging as a force in various radical Islamic movements and ideology as well as in other forms of nationalism. All seek to turn against the tide of modernity. The breakdown of traditional forms of life are all associated with the very structure of urban life and the ways that it patterns social relations and forms of authority are crucial in understanding the deeper mechanisms of these phenomena. There is no simple reduction of any single historical or cultural reality to the dynamics of space; but there is, I would argue, a close theoretical connection between forms of consciousness and the forms of association that exist in urban or non-urban areas. It is this theme, too, that any analysis of antiurbanism needs to address.

The chapters contained in this volume seek to study antiurbanism from a panoply of different perspectives. But in the end, the question that these various chapters seek to probe is the origins of antiurbanism itself, how it manifested itself in the change of residential patterns, in attitudes toward cities, or the ways that this theme manifests itself in literature and culture. Each of these studies moves us closer to a broader, more comprehensive understanding of antiurbanism not only as a voice of the antimodernist, but a more nuanced understanding of the ways that the human mind and culture react to the complexities of modernity and to the genuine pathologies and ills of urban life. In this sense, this book seeks to open up a new avenue for researching crucial questions in the ways that human beings organize their lives and the ways that the changing of human organization itself can lead to certain sets of ideas, political predilections, cultural and social attitudes, as well as different forms of cultural production. Antiurbanism therefore needs to be seen as a central area of study in the realm of social and political life, and it is with this sentiment that these studies are offered.

Notes

1. There are exceptions. Among the most important, see Morton White and Lucia White, *The Intellectual versus the City* (Cambridge, MA: Harvard University Press, 1962); Andrew Lees, *Cities Perceived: Urban Society in European and American Thought, 1820–1940* (New York: Columbia University Press, 1985); Richard Sennett, *Families against the City* (New York: Knopf, 1971); Richard Sennett, *The Fall of Public Man* (New York: Knopf, 1974); Lyn Lofland, *The Public Realm: Exploring the City's Quintessential Social Territory* (New York: Aldine de Gruyter, 1998).
2. For a discussion of the pastoral as opposed to the city, see Leo Marx, *The Machine in the Garden* (New York: Oxford University Press, 1964). For the literary dynamics of this in the ancient world, see Raymond Williams, *The Country and the City* (New York: Oxford University Press, 1973), 46–54. For more on the pastoral as a literary tradition, see Steven F. Walker, *A Cure for Love: A Generic Study of the Pastoral Idyll* (New York: Garland, 1987).
3. This is commonly a Marxian thesis: that the division between town and country emerges from a division between more complex economic forms and the division of labor in urban areas and precapitalist modes of production in rural ones. See Henri Lefebvre, *La pensée marxiste et le ville* (Paris: Casterman, 1972), 27–69.
4. For a discussion, see Isaac Kramnick, *Republicanism and Bourgeois Radicalism* (Ithaca: Cornell University Press, 1990), 177–183.

5. The classic study of this theme in American intellectual history is White's *The Intellectual versus the City* as well as N. J. Slabbert's essay, "Morton and Lucia White: The Anatomy of Antiurbanism," *Urban Land* 66 (7) (2007): 142–145.
6. Jean-Jacques Rousseau, *Emile, or on Education,* Allan Bloom (trans.) (New York: Basic Books, 1979), 234.
7. For excellent cultural histories of this trend in German history, see Andrew Lees, *Cities, Sin, and Social Reform in Imperial Germany* (Ann Arbor: University of Michigan Press, 2002) as well as Klaus Bergmann, *Agrarromantik und Grossstadtfeindschaft* (Marburg: Anton Hain Verlag, 1970).

PART 1

Theorizing Antiurbanism

CHAPTER ONE

What is Antiurbanism? A Theoretical Perspective

Michael J. Thompson

Introduction

Antiurbanism has a long lineage. Hatred of—or even ambivalence toward—the city or urban life seems a constant companion to the history of cities themselves. If we trace the long history of antiurbanism, we find a complex, varied phenomenon; one that changes as historical and political shifts make certain aspects of city either attractive or despised. In the end, antiurbanism is more than a mere hatred of city life. It is embedded in an overlapping series of economic, cultural, political, and sociological realities. It is a force that continues to have relevance in contemporary life, too, whether in terms of predicting or explaining political ideology or even mapping the variation of cultural habits and norms. Urban analysts tend to neglect the ways that nonurban areas and residents can manifest antiurban attitudes and behavior and they fail to grasp the importance of antiurban attitudes on political and cultural life. But in so doing, they also neglect some of the larger issues about the connection between space and consciousness, between ideas and location, and about the ways that other forms of social life—such as political values, religiosity, and so on—can be changed by nonurban environments. The commonplace view tends to be that nonurban areas are, in some way, those residues of the past; that they in some way are spaces where people have been untouched by cosmopolitan and modern ways of life.

The etymological origins of the word "urban" date back to the early seventeenth century. It meant "characteristic of city life," or more clearly, having the manners, habits, styles of expression of townspeople. The term was scarcely in use before the early nineteenth century, but it is interesting to note that the notion of the urban was always associated with ways of behavior, of thought, of the ways that actions of citizens within the city were distinct from those in the countryside. It is significant that the rise of urbanism as a way of life, as a new space for interaction and a new form of consciousness began to evolve in the mid- to late eighteenth century, the time when people like Jürgen Habermas have argued that a "public sphere" was beginning to emerge.[1] Urban life was associated with the new concept of the "public" that was being formed by a new, more fluid organization of society. It was a radical departure from the forms of social and cultural association that was structured by traditional bonds and rigid social roles. As a result, urban life became the crucible of modernity: the political movements, ideas, and associations that would remake the modern world had their origins in urban life and existence. It stands to reason that the contemporary shift from urban to suburban space will have some impact on this sociological reality.

On the surface these historical and cultural trends tell us little about the genesis of antiurban sentiments. Indeed, we could assume that it is a reaction to modernity, to the atomizing and disrupting effects of modern life that urbanism intensifies. This could explain the fear of urban centers as well as the antipathy toward them that we still see in contemporary life. But antiurbanism is more a more complex, more highly specified phenomenon than a reaction to modernity: it springs from the ways individuals intersubjectively form their social world, their moral concepts, and the social and political order that surround them. Antiurbanism is not simply hatred of urban life, of urban ways of thinking and acting—it is an anxiety response to the threatening dynamic of urban life, and in this sense, we need to look at the ways that antiurbanism is the expression of certain ways in which moral worldviews are formulated and reproduced. What I think is most important in producing and reproducing antiurban attitudes is the nature of the organization of space itself. Spatial organization is able to shape interactions between individuals thereby creating and shaping, through the process of intersubjectivity, the very internal dynamics of the ways individuals create an internal sense of moral order. At the level of social psychology, individuals must integrate themselves into their social world—urbanism and nonurban forms of life offer up two diametrically opposite ways of social integration, and they also have the ability to shape the subjective worldviews of individuals in radically different ways.

The focus and expressions of antiurbanism have changed throughout history. What is hated about the city in one period is not the same thing that motivates hatred of the city in another. But what unites all of these expressions is an emphatic rejection of urban life, of a linking of the urban with all that is in opposition to virtue, nature, truth. Most treatments of antiurbanism tend to be intellectual histories or cultural analyses.[2] Overwhelming attention has always been paid to processes of urbanization as well as to the "discontents" that spring from it. David Harvey has argued that "increasing urbanization makes the urban the primary level at which individuals now experience, live out, and react to the totality of transformations and structures in the world around them."[3] But even as urbanism has been the spatial component to the process of modernization, it would be wrong to view it as the "primary level" within which people experience daily life. Mass suburbanization in the United States, for example, has signaled a reverse trend in the overall historical process of urbanization and has itself led to a renewal of an American antiurban ethos.[4] Elsewhere, we find the processes of urbanization and modernization giving rise to an antiurban impulse as well. Even more, as the city continues to change its cultural and economic role in the developed world, the rift between urban and nonurban continues to deepen. Categories such as gender, sexuality, and other dimensions of personal lifestyle are reworked and questioned in the city that gives rise to an opposition to them and to urban life more broadly within nonurban areas. At the same time, decayed cities—especially in the United States—become symbols of racial stigma and societal decay. The city becomes the emblem for what is worst in society.

But antiurbanism is not simply a dialectic to urbanism, it rests on certain forms of association, certain cultural ways of understanding of society and self. Just as urbanism ushers in modern forms of life and consciousness, antiurbanism is a rejection of those social forces; it is a tradition of thought, to be sure, but also—and perhaps more importantly—a product of a tension between different forms of life and the kinds of association to which they give rise. These forms of association produce different fields of social relations and ways of understanding by patterning certain kinds of relations that then affect the structure of group and individual consciousness. Antiurbanism is a reaction to the forms of association that in fact make urbanism possible; it is a reaction against certain ways of association, but this reaction stems from the need to protect moral order, to shield oneself from an openness of association even as it is also many times a desire to flee crime, and lower quality of life. In this sense, the referent of antiurban attitudes needs also to be expanded into the social psychological realm and this has deep moral and political consequences.

The more general hypothesis that will guide my investigation here is that antiurbanism is grounded in the ways that individuals create and recreate their sense of self, of a normative order within which they live. Antiurban sentiments are a response to the ways that this order is created and maintained since, I will contend here, urbanism is itself not simply a matter of location, of cities themselves, but more importantly, it is also a set of ways in which people interact, think, and act that are different from nonurban areas. Within urbanism, one encounters not only difference, the "other," and so on. One also encounters new ways of interacting, new threats to the ingrained and internalized moral-political belief systems that individuals possess. They are forced to rethink their assumptions about others, about certain value systems, about their own conception of self, and so on. In this sense, antiurbanism is a more pervasive reality than previously thought and, I will show, is not something restricted to nonurban areas alone, but also a feature of many urban areas as well. Antiurbanism is a defensive response to a perceived assault on a moral self that seeks to protect a conception of the world and its place within it at all costs. It seeks this protection because it fears revising the worldview it has constructed for itself; one that shelters it from the possible disruptions and realities of a world in flux, a world where individual autonomy requires that self to encounter what is new, where it is dared *to be autonomous, to be an individual* in the truest sense. In this way, antiurbanism is closely tied to the realm of politics and to the ways that spatial forms shape social relations that in turn shape individual (e.g., moral) consciousness. This is the first step in constructing a theory of antiurbanism.

Space and Society

The material foundations of antiurbanism are found in the ways that social space is structured and in the ways that social relations are shaped by this restructuring. Antiurbanism is, then, a product not only of certain political and economic interests, it is also a product of a certain *form of consciousness*—a form of consciousness that is a function of other social forces (economic, political, etc.). Much of this analysis depends on the theories of association and space developed by social theory since the late nineteenth century. Antiurbanism is not simply an ideology, it is produced by concrete forces that lie at the basis of social organization and the ways that space is patterned and the ways that this patterning itself is a function of deeper, more structural economic and social forces. Theorizing antiurbanism therefore requires the insight that social practices are products of social structure and the ways that social structure affect consciousness.

At the heart of the relation between space and society lies the issue of association and the ways that space can either impede or encourage certain forms of association.[5] To this must be added the insight that human thought and action is in fact "embedded" in space, and this means that the very intersubjective nature of human consciousness and thought can be affected by the nature of the relationship between association and space. The divide between urban and nonurban spaces has been an issue of study for quite some time, but what needs to be grasped is the ways in which space, interaction, and ideology come together to produce certain forms of life and ways of thinking and acting that reaffirm and support those forms of life.[6] Spatial configurations are themselves subject to other structural contingencies or forces. Economic and political movements and dynamics can create, destroy, or shape spatial configurations that in turn impact the ways that association and interaction take place. It is well known that association within suburban areas, for example, tends to be more diffuse than in urban ones, or that association tends to be more predictable and stable, less confrontational, and so on in rural ones—and this can have cascading consequences on consciousness and culture, explaining differences in cultural, political, and social psychological patterns of behavior and attitudinal formation.[7] The connection between space, association, consciousness, and ideology is therefore at the crux of the analysis of antiurbanism.

In classical social theory it was Ferdinand Tönnies' famous distinction between *Gemeinschaft* and *Geselleschaft* that has remained one of the more robust theoretical formulations of this divide. The real insight of Tönnies—and the other social theorists that followed in his path—was the distinction between the ways that communal forms of life fostered different kinds of social bonds between individuals than did more urbanized forms of life. More specifically, Tönnies' distinction was not between different forms of spatial arrangements but rather between the different *ways that individuals related to each other.*[8] Tönnies was concerned with the distinction between individualistic and rationalistic models of association as opposed to ones grounded in traditional mores, folkways, religion, and a sense of common values and culture. His insights would be developed by numerous theorists, but his fundamental insight that the kinds of relations individuals have with one another was crucial in understanding broader, shared social attitudes, values, and so on remains salient. And although he was clear that this was not something that was tied to space, it can be argued that the different ways space is constructed leads to different modes of interaction and relation—the shopping mall versus the town square, just as the density of urban networks as opposed to the relatively sparse atomism of suburban life, and so on.[9] The point is that explaining antiurban phenomena is crucially linked to understanding the complex interrelations between space, association, and their effects on social and individual life.

Tönnies was able to see that ways that different forms of association also impacted different cultural logics of communities. It was Georg Simmel's insights which made the crucial connection between the outer, social context and the inner, "mental life" (*Geistesleben*) of the individual. Simmel's emphasis was on the ways individuals within cities, or the ways that the "personality accommodates itself in the adjustments to external forces."[10] The crucial insight here was the in the way that the process of individualization was connected to the impersonal, external, social processes of society. The mental and "psychic" domain was therefore linked to the outer social domain, and this provides us with a crucial aspect to understanding antiurbanism. Although Simmel's attention was placed on urban life, it is important to see that the development of the personality in nonurban contexts can give rise, through a series of processes, to distinct attitudes and beliefs that can produce antiurban sentiments and behaviors.

Antiurbanism and Mental Life

The link between the material/spatial organization of society and the experiential and mental organization of the subjects that inhabit it is therefore the place where an understanding of antiurbanism can be most clearly understood. Unlike some theorists of the decline of public life or of civil society and "social capital,"[11] I think that the spatial organization of nonurban life is a central explanatory variable leading to distinct attitudes and sensibilities that shape certain cognitive (specifically social) views toward urban areas. Antiurbanism is produced and reproduced by the ways that individuals are shaped by their social environment, and by this I mean the various ways that kinds of interaction can shape epistemological habits of thought and ways of interpreting the world. By frustrating interaction between individuals, intersubjectivity becomes hampered leading to certain consequences reflecting themselves in the sphere of political values and belief systems. In this sense, the connection between the organization of space and the organization of social experience become more intimately related.

But in addition to the frustration of social interaction, there is also within nonurban life a differentiation of *kinds* of citizens: the division between urban and nonurban space also constitutes differences between race, class, and ethnicity, not to mention different ideological variety as well. As a result, it is not simply in the frustration of interaction itself that can explain the emergence and maintenance of conservative political and social values in nonurban areas, but also with whom people interact. In this sense, the cognitive capacity for political life and practices is affected by interaction, but so is the way that individuals conceive of themselves. At a deeper level, nonurban space has a deeper impact on the ways that individuals shape their moral and political views about the world and provide them with little capacity or opportunity to question them. In this sense, social interaction itself cannot be viewed outside of the context of with whom one is interacting. Attitudes and values are shaped within the context of an interaction, and the possibilities for different kinds of interaction is itself shaped by spatial structure.

In this respect, we must distinguish between the realm of spatial organization on the one hand and the realm of the organization of experience on the other. Spatial organization can affect mental experience by either enabling or setting up barriers to association. The more intersubjective relations individuals encounter, the more developed certain cognitive capacities for questioning dominant norms and value systems can be nourished—not only the more common practices of discussion and debate, but actual cognitive capacities that enable individuals to adopt alternate or opposing perspectives and expand their horizon of moral and political belief. From a theoretical point of view, each of these spheres is concentrically organized so that they act as layers to the process of individual conscious-formation. The relation between the outer environment of the individual and their inner, psychological life has been probed before[12] but has not been able to bring these different spheres together to theorize political attitudes and values, let alone provide a theory for the geographic pattern of voting behavior. More to the point, I think that each of these different spheres of social and psychological reality needs to be seen as fluidly affecting each other. The core insight is therefore that the relation between space and consciousness can be seen to arise from the ways that space structures social relations between individuals as well as the ways that space communicates to them as well.

The crucial aspect of theorizing the relation between space and consciousness is to link the spatial-structural realm with the phenomenological realm. In this respect, we must see how it is that spatial structures affect and shape ways of seeing and feeling and thinking. Individuals formulate their ideas about the world based on the socialization of their experiences. Put another way, they learn ways of "framing" their perceptual world that "provide background understanding for events that incorporate the will, aim, and controlling effort of an intelligence, a live agency."[13] In this sense, it is more important to focus on the link between the ways that the structure of an individual's spatial environment leads to certain ways of "framing" the world than on the emphasis of social practices by themselves[14] since the act of framing, of the way one organizes one's perception of the world, acts as a legitimating factor not only on everyday life practices but also ideology as a whole. For Goffman, a frame is a "schemata of interpretation," and in this sense, it provides us with a way of making sense of the world and the ways a subject sees him or herself within this context. These frames are therefore crucial to understand the ways that individuals organize the experience of their world, and shape their moral and political conceptions.

Goffman's insight brings us closer to a theory of antiurbanism since he is able to provide a theory of the ways individuals experience their perceptual world through certain "grids" or "basic frameworks of understanding available in our society for making sense out of events."[15] Spatial structures serve as means by which these frames are produced and maintained since certain forms of interaction become limited by the ways that space is organized. Social interaction is a primary way in which the individuals' frames are constructed and maintained since it is through them that they gain access to ways of thinking about as well as within the world. It is through social interaction that individuals learn norms, form conceptual schemas about moral categories, and so on. Space therefore is a formal category that can shape interaction and association and thereby create frames that individuals internalize.

The first way in which this operates is through "distanciation" that is a way of measuring the extent to which individuals can overcome spatial distance in order to engage in interaction.[16] Forms of suburban life tend to create barriers to overcoming distanciation: reliance on automobiles, the lack of public space, as well as the proliferation of detached, single-family homes, all create barriers to social interaction.[17] Second, there is the issue of *homogeneity*. Much of nonurban life depends upon the separation of communities and, despite some change in the general trend, a separation of class and race as well.[18] There is also the lack of ideological diversity that one is more likely to encounter in urban areas. What this means is that the spectrum of diversity—whether based on ethnicity, race, class, gender, or ideology—that individuals have access to through interpersonal interaction and association becomes narrow and limited. Frames are therefore formed that tend to exclude or limit certain moral categories, ideas, and beliefs as well as reproduce certain hierarchical forms of social life having distinct political outcomes in terms of the political attitudes and preferences individuals possess.

Experiential frames are derived from such structural issues because forms of interaction become limited and predictable. One encounters others within public life with decreasing frequency and social relations themselves become transformed as the traditional forms of social interaction are supplanted by increasingly mediated forms of interaction: economic relations lose personality once in shopping malls and supermarkets; car transportation prevents social interaction during commuting and traveling; and the lack of public space constrains other forms of social and public interaction.[19] This mediates consciousness by limiting intersubjective relations to the sphere of the family and the workplace, both places where individuals can find themselves constrained by predefined, functional hierarchical relations.

Association therefore provides more opportunity for individual selves to move beyond such constrained forms of self thereby affecting the way one's social consciousness is shaped. The genesis, reproduction, or at least the predominance of conservative values in nonurban areas can therefore be traced to two distinct ways that the self can be produced in nonurban space: (i) a self that does not question or seek to conflict with other views (political, moral, etc.); and as a result (ii) a self that seeks to maintain and protect predefined forms of moral order.[20] The limiting of intersubjective relations also has the effect of causing an alienation from public concerns and public forms of thinking and acting even as higher levels of association that lack intersubjective heterogeneity can equally lead to a reinforcement of certain norms and belief systems that can also encourage certain kinds of conservative thinking (e.g., an emphasis on family values, conservative views on race, gender, and ethnicity, etc.). Hence, we can differentiate between three different "frames," or forms of consciousness affected by the structure of social space: interpersonal, reflexive, and public.

Interpersonal consciousness reflects the ways that individuals relate to others and the way they think of themselves in relation to others. Ways of talking, arguing, coming to some sort of mutual understanding between individuals requires, at its base, the interaction between individuals in a rational form. This is contrasted to more intimate forms of interpersonal interaction such as between family members that are structured and circumscribed by custom and emotive forms of relations.[21] Going back to the categories of distanciation and homogeneity, we see that this frame becomes accustomed to limited forms of interaction due to the dominance of the private sphere over the public realm and its forms of diversity and "thickness" of association and interaction.[22] An excessive lack of association can lead to distorted forms of interpersonal consciousness where the individual's conception of self is defined only in relation to what he or she is exposed to in terms of the most immediate social contexts: such as the family, certain communal belief systems, and institutional forms of authority such as the workplace or school. More exposure to association can force the individual to test certain belief systems and closely held opinions and therefore encourage a more rational and less purely subjective orientation of belief and thought.

Reflexive consciousness refers to the relation an individual has toward himself as a member of a community. Interpersonal frames have the capacity to overlap with and structure personal frames in the sense that interpersonal relations have an effect on the ways that individuals organize their experiences and therefore form notions of the self. These include their self-perceived obligations in terms of politics but they also concern certain moral conceptions that refer to the relations individuals deem to have toward one another. In this sense, individuals with certain values or points of view can be seen as operating outside a predefined communal value or belief system. Reflexively, the individual situates himself within a group but it also makes it possible for him to adopt the perspective of others *within* himself, "to turn an experience back on himself."[23] In this sense, the form of reflexive consciousness enables a capacity within individual consciousness to "take the attitude of the other in our various life-processes."[24]

Finally, *public* consciousness can be defined as the ways in which individuals approach their political world, whether at the local level or at the more "abstract" or macro level. It denotes a frame of consciousness that enables the individual to organize an interpretation of the political world and the way he fits into it. Within this frame, the individual is aware of effects of certain realities on other people outside of himself, he is able to make judgments about the organization of the community as a whole. Public frames therefore put shape to the ways that individuals see themselves as part of a broader community. Individuals with more developed frames of public consciousness will be more likely to participate in public life rather than avoid it. They will be more likely to be aware of the broader issues and political realities which affect the community as a whole rather than view it more narrowly through self-interest or in individualistic terms. The frame of public consciousness is therefore a crucial layer in democratic forms of life. It requires that the previous two frames also be well developed since an individual must also have the cognitive capacity: (1) to interact with others and be able to exchange and process the information derived from interpersonal encounters (interpersonal frame); and (2) to distance oneself from one's own closely held views and attitudes and be able to see his own ideological views from a third-person perspective and examine them critically from that vantage point (reflexive frame).

The frame of public consciousness is also crucial since it is here that individuals are able to situate their ideas or sensibilities concerning different political issues and interests. The reproduction of certain value and belief systems concerning race, ethnicity, class, gender, and so on are facilitated once we see that there becomes less and less opportunity for individuals to interact with "others" and have exposure to other forms of life and ways of thinking about, or framing, the world. But even more, it provides us with a crucial insight: namely that the ways that individuals constitute their external world is dependent upon the forms of association within which they evolve.[25] That although these attitudes ultimately may be subjective in nature, that they are created by ontologically objective social processes and that these social processes are not in any way arbitrary but are in fact determined by the nature of spatial organization. In this sense, antiurban attitudes can be traced, I think, to very determined variables: namely to the ways in which the variations in spatial organization are capable of shaping intersubjective practices which, in turn, shape cognitive processes of individuals. These mental processes are responsible for generating the a priori understanding of the social, moral, and, by extension, political world for the individual in the sense that these forms of consciousness have the capacity to shape what the individuals see as right and wrong, as comfortable as opposed to fearful. It is to this domain of antiurbanism to which we must now turn.

Antiurbanism and Political Values

A key aspect of shaping the internal lifeworld of individuals remains the ways in which individuals become integrated into the social order around them. The ways that individuals integrate themselves into a broader social order is largely derived from the dynamics of the socialization process that structures his conceptions of the social world. But even more, it structures the normative moral order for the individual: it provides a priori categories for what are considered right and wrong, for what is desirable and what is to be shunned. It structures normative behavior as well as a conception of a morally "correct" social order and one's place within that order. In this sense, the political elements of antiurbanism can begin to be sensed since moral ideas of right-wrongness are crucial to a beginning of political conceptions. Particularly in understanding the ways that political ideology differs from urban to suburban and rural areas, this issue becomes central in our understanding of the ways that antiurbanism intersects with the moral-political realm.

In historical epochs previous to modernity, the city was a place not only of corruption and exploitation, it was also a place where ingrained conceptions of their social and, more importantly, moral order. Antiurbanism is a deeply political phenomenon. It affects voting behavior, conceptions of political ideology. Antiurbanism is generally associated with conservatism, but not only because of the insight of traditional social theory: for example, that it is a reaction to modernization and the division of labor. Rather, I think it derives from the analysis I presented above: namely, that nonurban forms of life can restrict more robust forms of association *and* a heterogeneous intersubjective environment. Thicker forms of association develop within the individual different frames of social consciousness, as I argued above; but heterogeneity defined as moral and interpersonal diversity of subjects, are both crucial in understanding how antiurbanism becomes a causal variable in the emergence of conservative moral and political views.

But how does this actually work itself out? How is it that nonurban forms of life create not only a hatred of urbanism, but also shape more conservative values within those individuals?[26] The explanation is bivariate since prolonged social and public interpersonal detachment characteristic of certain forms of nonurban life (i.e., suburbs) tends to lead to avoidant-attachment feelings since these individuals tend to see the world as being uncaring and cold, and a preference for hierarchical forms of life become comforting.[27] The social world and the moral order of their immediate environment therefore become crucial as a shield from difference and change that they see as threatening. But equally with more rural, *Gemeinschaftliche* forms of life that incite anxious-attachment feelings of needing to protect the individual from the dangers of whatever exists outside of their known environment.[28] Urbanism therefore becomes a crucial object of fear since within suburban and more rural contexts, individuals will tend to view urbanism as threatening to the moral order that they have internalized. Willingness to question, examine, and especially revise that moral order and belief system therefore becomes less likely not only because these individuals will tend to avoid those places where they will confront threats to their worldviews, but also because nonurban space offers little in the way of interaction and heterogeneity (whether ideological or otherwise) to serve as oppositional to ingrained belief systems.[29] The need to ensure social conformity can therefore lead to prejudicial opinions about those who are "different" and threaten perceived moral and social cohesion, leading not only to conservative political attitudes and beliefs, but even in more extreme cases, toward right-wing authoritarian attitudes as well.[30] Indeed, social conformity therefore becomes not simply an option for each to decide for him or herself, but rather becomes reflected into political views about the world as a whole.

In this sense, political values, sensibilities, and attitudes can be seen to spring from forms of consciousness—that is, ways in which individuals frame their political, social, moral, and cultural environment—that are shaped by the forms of association and intersubjective relations that constitute their social world. Seen in this way, spatial forms that frustrate association through distanciation, spatial separateness, lack of public space, lack of communal heterogeneity, and so on will tend to produce individuals with less experience with associational life that itself can retard or prevent the development of forms of public reasoning. The frames produced within these social contexts tend to be shaped more by immediate family ties, small groups of familiar groups where familiarity and homogeneity tend to be preferred. Space acts as a means by which these forms of life are themselves shaped: families become more insular, individuals undersocialized, all with less public space and less opportunities not only for civic association, but also—and I think more importantly—without a culture of interaction with ideas and values different from their own. Space affects the social psychological domain by shaping forms of interpersonal interaction and thereby shaping values and attitudes at a much deeper level within the individual leading to certain conservative political predispositions.

It is no surprise, then, that we see antiurbanism at its strongest when urbanism begins to emerge in any historical period since it affects a major disruption in the ways that social relations are structured and conceived as well as the ways they are legitimated. Key to the political aspect of antiurbanism is in the ways that human intelligence develops through public association to critique ways of life and moral precepts that are predominant in society.[31] The history of American politics, for example, is replete with movements against urban life, some of them coming into maturity in the late twentieth century just as it is replete with progressive movements coming out of the urban context. The political implications of antiurbanism stem from the ways that culture itself can affect political interests. The antiurban sentiment is far from monolithic. Empirical studies have often found acute breaks in political ideology and culture between urban and rural areas. If the general theory that there is a relation between individual consciousness, of ideology, of patterns of thought on the one hand and spatial configuration on the other, then antiurbanism will continue to be a relevant frame of analysis as suburban trends continue in the developed world and urban trends in the developing world.

The relation between conservative politics and antiurbanism is a staple since the patterning and ordering of social relations and therefore of social power that becomes more manageable, more predictable within nonurban environments. Fear of the "other," the closing off of public space in favor of private space, the new emphasis on the family as a center of everyday life, all enter into the complex story of how antiurbanism is tied with conservative political ideas and dispositions. In American conservative political discourse, the relation between conservative ideology and "family values" has had a distinct antiurban cast since it emphasized the nuclear family and its sheltering from the perceived fluidity and impermanence of public life, especially those aspects of family life that were not only dangerous in real terms, but were also "dangerous" in the sense that it threatened the moral order of its participants.[32] Even more, the urban once again takes on the cultural significance of sin, secularism, of decadence, and moral corruption. Even outside of the United States we see a similar patterning of attitudes. Antiurbanism therefore results from a complex of factors. Different social and cultural forces may make the antiurban impulse stronger in some places and times than in others, but in the end, we contend that antiurbanism ultimately leads to political consequences that have the effect of transforming, shaping, and redirecting social development in certain ways. Antiurbanism ought no longer be seen simply as the underside of development and modernity, but as a permanent feature of the complexities which arise from modern life. The tension—at the level of the sociological as well as at the psychological—that arises from the emergence and development of modern forms of life and consciousness is at the heart of the antiurban mindset.

American Suburbs, Antiurbanism, and Democratic Sensibilities

As a brief case study, it might be interesting to look examine the thesis that suburban life breaks down democratic culture and sensibilities. Suburbs are, in many ways, the result of antiurbanism in thought and practice. Peopled by those fleeing the city, by those wanted to abandon urban life in favor of security, property ownership, and the pathologies that had always characterized urban life, suburbanism provides us with an interesting place to study the emergence of antiurban sentiment and its causes. Suburban life also contains some of the more important elements of the above theoretical analysis: the erosion of associational life, on the one hand, and the lack of diversity—ethnic, racial, ideological, and so on. What I want to get at here is the ways political sensibilities can be formed and reproduced. More specifically, suburbs provide a spatial pattern of social life that actively erodes the interactive social foundations of everyday life thereby, in time, leading to an erosion of democratic sensibilities and democratic forms of life.[33]

Whereas urban environments are characterized by diversity, a density of social interaction, and a constant exposure to difference and newness capable of spawning a sense of openness and constant sense of newness, and ways of innovating and exploring what Georg Simmel referred to as "the technique of life," suburban life is characterized by an isolation from those very activities and external forces. It is defined by the fact that one can isolate oneself from community; it is the spatial manifestation of the liberal political and cultural utopia: to be able to separate public and private at one's own whim and be able to live unencumbered by the various obligations of public and social life. Suburbanism was seen as an escape: an escape from the conditions of urban life, from the necessity of cooperation and interaction, and the desire—only realizable on a mass scale during the affluence of post–World War II economic expansion in the United States—to avoid difference, or, as Lewis Mumford wrote in his *The City in History* in 1961, "the ultimate effect of the suburban escape in our time is, ironically, a low-grade uniform environment from which escape is impossible."[34]

Less than a decade later in his analysis of the urban-suburban situation in the United States, Richard Sennett wrote about the rise of a "new puritanism" where family life became the focal point of suburban life, a desire to intensify familial relations through the simplification of social environment was sought out.[35] Sennett was simple and direct in his analysis arguing that "the desire of people beyond the life of economic scarcity is to live in a functionally separated, internally homogenous environment; that is the crux of the matter."[36] For Sennett, as with Mumford, suburbanization represented—albeit in different ways—an erosion of diverse communities, and the emergence of the possibility for individual isolation within the framework of a uniformly homogenous society. For them, as with some other critics of the time, this was leading to an aimless and indeed empty social and cultural life that was something wholly new in modern life and individual consciousness.

Suburbs do several things that foster and maintain antiurban sentiments that can have deleterious effects on democratic (e.g., more tolerant, inclusive, public-minded, etc.) political sensibilities or values. First, they reduce publicly interpersonal association through spatial design. Suburbs frustrate public forms of social interaction in several ways: by the absence of public space, by the relative, or in many instances absolute, restriction of ethnic/racial and class diversity, making them largely homogenous. In this sense, individuals' ability to expand their moral horizons becomes limited, leading to a fear of interaction with those who are unfamiliar, leading to a fear of urban environments and the perceived threats that they engender.[37] This fear encourages certain social psychological impulses within individuals since the "ego-defensive" attitudes guarding against those ideas that oppose or disrupt the moral order understood by subjects and places where such threats are common (e.g., urban areas) gives rise to conservative values and also predispositions to authoritarianism.[38] This occurs because the moral safety offered by suburban life is actually quite brittle, since the network of weak social bonds forces individuals inward thereby making such individuals more vulnerable to perceived "ego-threats."

Second, suburbs reduce interpersonal conflict thereby further enhancing social anomie and undermining civic life.[39] In addition to furthering the anomic effects of suburban life, it also deprives individuals of certain social psychological capacities of public life, of acting and thinking publicly. When an individual negotiates conflict with another, he develops a capacity of arguing, a capacity to articulate formal arguments, and so on. This is, of course, an ideal case, but it nevertheless can lead to deeper levels of trust if conflicts are resolved mutually rather than through institutions. Conflict need not be understood narrowly as clashing self-interests, it can and should also be understood much more broadly as a conflict between different moral and political views. Argument about public policy, about politics more generally becomes negated by the relative lack of interaction and moral minimalism of suburban life. Individuals are more likely to have weak ties to other individuals and keep moral and political issues personal seeking to avoid conflict.[40] This has the effect of limiting the capacity to see politics as a shared commitment and even more, it deprives individuals of crucial cognitive skills when it comes to negotiating moral-political issues with others and, even more importantly, revising one's own personal understanding and system of moral-political beliefs. Fear of difference, of change, of disruption begins to set in. In place of a fluid, dynamic public life that leads to a constant revision and rethinking of political and moral values, individuals within suburban contexts are closed within the sphere of privacy.

Urbanism therefore becomes an object of fear, mainly because the moral order that individuals construct for themselves becomes crystallized. Urbanism is a place where these notions of moral order are put into question, where the likelihood that they can be threatened increases. Threats to that order become severe, and this leads to patterns of political belief and moral conceptions about the world and about cities and urban areas more specifically. The anomic life of much of suburban life creates a culture where individuals—through a relative lack of interaction and interpersonal conflict avoidance—are undersocialized and, in many ways, less inclusive and less open to different political and moral perspectives. Democratic ways of life are generally defined by our capacity to call into question our prevailing views and beliefs; to argue their validity or to see their invalidity through some kind of public dialogue. It therefore becomes much easier to see how suburban setting can lead to conservative political views: those who seek to emphasize family values (the one refuge for human interaction in suburban life considering the relative absence of public space and public life), as well as ideals of individual protection for one's privacy at the expense of public goals. Suburbanism erodes civic consciousness not only by undermining social association in itself a la Robert Putnam's argument, it more importantly shapes individual consciousness itself, thereby structuring the ways that individuals form and process moral and political conceptions about their world. It provides a cognitive mapping—a "frame" in Goffman's sense—of the objective world.

But there is more. To this lack of conflict, lack of associational life, lack of moral and political cognition and self-critique also must be added an increasing dependence on outside institutions. The second part of this argument therefore consists of the fact that the suburbs mesh with the particular changes in economic and social life that have slowly occurred throughout the postwar period but began to accelerate during the closing decades of the twentieth century. Namely, the fact that there has been a rise in economic inequality that has given rise to increased working hours as well as increased consumption. This has had the effect of confining most suburban Americans more and more to two institutions that are, by their nature in American life, largely antidemocratic: the workplace and the family. The lack of leisure time, combined with the problems that they are indicative of suburban life (namely that of a lack of density, diversity, publicity) give rise to what can be called a "new provincialism."

The political impact of this, however, is severe. By leading to an insulated form of individualism that eschews cultural difference, it has led to the increased isolation of different groups. Racial and class groups are more segregated between spatial location—that is, urban centers and their suburban peripheries—leading to what Douglas Massey and Nancy Denton have termed an "American apartheid"[41] itself leading to deep impacts on the nature and practice of democratic life within suburbs.[42] The increased emphasis on individual and family life has led to a new provincialism that becomes ignorant of other cultures even as the world becomes increasingly global and interdependent in nature. Urban areas provide increased access to newer, denser social networks and expose their inhabitants to difference and modern urban life tends to have more liberal, more tolerant political values as opposed to suburban and rural areas. Historically, this has always been the case, and new research needs to be done into the deeper dynamics of this aspect of urbanism and its implications for modern political life.

Politics, in other words, should be seen as possessing what could be called a *spatial embeddedness*. And, by implication, urban and suburban locations also can shape political ideas, values, and voting behavior. The space of everyday life has much to do with the ways that people think about their social and political environment. The lack—or the inconvenient placing—of public spaces, the architectural banality of public buildings, the relative residential separation and isolation of suburban housing, and the reliance on private car transportation systems, all contribute to an erosion of the public sphere, an indifference to broader political concerns that lie outside of the most immediate issues of communal and individual interests (e.g., those that surround concerns for one's own property value and taxes), and a reinforcing of atomistic individualism, or what thinkers like Robert Putnam have described as ever-lessening "social capital." Culturally, suburbs are largely, if not entirely, cut off from cultural institutions such as museums, concert halls, theaters, universities, and the like that enable a new exposure to new sensibilities and to cultivate them to a degree not possible within the confines of suburban life.

With the outer domains of social and cultural life largely absent, suburban life revolves around the institution of the family and the instrumental pursuits of property (specifically home ownership). What I have above called the "new provincialism" has severe effects on critical political reflection and participation. There are two main reasons for this. First, there is the problem of the limitations of self-interest in democratic politics. Suburban life is the spatiocultural ideal of the normative assumptions of classical liberalism. On the one hand, the ideal of private existence separate from the public sphere was something that was supposed to allow individual liberty to fulfill the dictates of one's own life choices, or modus vivendi. Freed from the restrictions of tradition, servitude to others, and/or religious dictates, the individual was to have sovereign reign over his existence, the means to this existence, and the particular life path that he chose for himself. The only limiting factor was that these choices and actions were not to interfere with others—the social contract was to create a sphere of action where others would not be harmed by your particular freedom. But under conditions of modern life, this has become an aggressively atomistic doctrine that has eroded other forms of social solidarity and communal relations that once were considered—even by most theorists of classical liberalism—assumed. The pursuit of self-interest at the expense of most social and public aims and goals is the hallmark of modern American life, but it is one that has been intensified, if not made explicitly possible, by spatial embeddedness that suburban life offers. With the very nature of the public now a mere abstraction, participation in it becomes equally so.

The second way that suburban life has had the effect of eroding democratic life is in the way that this new provincialism has laid out a sterile notion of everyday life and existence that has had the effect of the acceptance of some of the most undemocratic forms of life in modern, advanced societies. With little access to a vibrant public sphere or cultural institutions, and lacking a communal style of life that seeks out such institutions and activities, suburban life throws the individual onto two institutions that structure everyday life: the workplace and the family. Both institutions—especially, as Sennett's work points out, the family—are largely hierarchical and antidemocratic in nature. Growth outside of these two institutions becomes difficult within suburban space since the very physical distance from more culturally concentrated urban centers makes access to alternative forms of life and activity difficult. The family becomes dominant institution outside of the workplace that, itself, is highly antidemocratic and stifling.[43] The economics of suburban life—necessitating huge debt to afford expensive mortgage costs—therefore becomes dependent on the demands of the workplace and its affects upon individual consciousness: of institutionalized hierarchy, a decrease in economic security, expanded working hours, and so on. With deteriorating benefits of vacation and time away from work, people are more tied to their locations than ever before, and their entrapment in their homes and work life further alienates them from public life and civic affairs.

The issue of the move from community to that of atomism is also something that is intensified by suburban life. This is also not a terribly new insight.[44] Louis Wirth's seminal analysis of city and of community was premised in the definition of community as interdependence and communication. He argued that as this began to break down "we create interests units."[45] In other words, the breakdown of communicative, intersubjective social life breeds self-interest at the expense of public interest. The democratic element of local life therefore breaks down as well as the personal is premised over the public. The classic notion of the citizen that dates back to Aristotle's *Politics* that defined the good citizen as one who put the public good over the good of the minority or the one, vanishes. Since suburbs are based on segmented private property units by design, political concerns are increasingly circumscribed by individual property as the prime mover of political interest, sealing off larger social problems of inequality, segregation, and local funding for public goods.[46] The lack or even complete absence of public space within suburban areas is central to this erosion of broader political life, and this was, in part, by design. Dolores Hayden has shown how postwar suburbs "were deliberately planned to maximize consumption of mass-produced goods and minimize the responsibility of the developers to create public space and services."[47]

Conclusion

Antiurbanism is a phenomenon deeply rooted in the ways that people live their lives: in their spatial, communal forms of organization within which they live, and the ways that these outer forms of reality shape their process of individualization. It is a phenomenon that occurs at the nexus of the objective and subjective worlds, and from this process, certain moral and political values, beliefs, and practices emanate. My hypothesis in this chapter was that antiurbanism is more than an intellectual tradition, more than a mere response to modernity, to the degeneracy of cities and urban life. It is also, and more importantly, an enduring feature of many individuals and can account—at least in the United States, as an example—for the growing cultural divide in American politics between liberal metropolitan centers and more conservative suburban, exurban, and rural ones. Antiurbanism has political consequences because it is a phenomenon caught within the ways that individuals shape their conceptions of the world and of themselves. For human beings, this means forming their attitudes and values from moral concepts—moral concepts that are themselves formed sociologically: by the kinds of interaction and intersubjectivity to which an individual is exposed.

But it would also be a mistake to assume that antiurbanism is something that occurs only in nonurban areas. If my hypothesis is correct, then it is easy to see that antiurbanism can also be found within urban areas. This may seem contradictory, but not if we examine many of the trends that gentrifying areas typically encounter, namely a "suburbanizing effect" with respect to the ways that new inhabitants seek to transform the neighborhoods they inhabit.[48] Many of these changes—the movement of poorer residents out of the neighborhood, the proliferation of shops, and so on—all attest to a certain urban-antiurbanism: an attempt to create some form of communal homogeneity, personal safety, and avoidance of public life and association. Although not as intense as in nonurban areas, it is an observable phenomenon. But even more, the creation of certain enclaves within urban areas that shield those communities from the outer urban world are another example of antiurbanism in the urban context. What these neighborhoods suggest is that urbanism is a consistent threat to the private self, that they are in many crucial ways, expressions of ways that a public culture has broken down within cities themselves. It is therefore a mistake to make a crude distinction between the city and the country when it comes to antiurbanism.

There are many other elements to antiurbanism as a mindset and even as social policy. But the real essence of the story is that social researches need to be more aware of the spatial dimensions of social and political life as well as the ways in which space can deeply affect the consciousness of individuals. And as other parts of the globe continue to modernize, industrialize, and urbanize, antiurbanism will no doubt continue to present itself as a crucial social, political, and cultural force in the time to come. But even more importantly, the mechanics of antiurbanism should give us a glimpse into the deeper valences a theory that has been elaborated for over half a century now in social and political theory: namely that a breakdown of public culture and public sphere has been underway. That a crucial aspect of modernity has been the breakdown of social groups, civic associations, and political participation. In many ways, antiurbanism—at least in the way I have approached it here—should give us a deeper insight into the ways that space affects the psyche of individuals and the nature of political culture.

Notes

1. Jürgen Habermas, *The Structural Transformation of the Public Sphere* (Cambridge, MA: MIT Press, 1989).
2. For classic examples, see Morton and Lucia Wright, *The Intellectual versus the City* (Cambridge, MA: Harvard University Press, 1962) and Raymond Williams, *The Country and the City* (New York: Oxford University Press, 1973). For extension of the analysis of Williams, see G. MacLean, D. Landry, and J. Ward, eds., *The Country and the City Revisited* (New York: Cambridge University Press, 1999).
3. David Harvey, *Consciousness and the Urban Experience* (Baltimore: Johns Hopkins University Press, 1985), 251. For a discussion of Harvey's work in the above context, see Ira Katznelson, *Marxism and the City* (New York: Oxford University Press, 1993), 116–140.
4. For a discussion of this theme, see Robert Beauregard, *When America Became Suburban* (Minneapolis, MN: University of Minnesota Press, 2006), 70–100; and Daniel Lazare, *America's Undeclared War* (New York: Harcourt, 2001), 187–211. For a more developed discussion of the relation between suburban life and the changing dynamics of culture between urban and suburban areas, see Lizabeth Cohen, *A Consumer's Republic* (New York: Knopf, 2003).
5. The most interesting work on the ways that space limit or shape the context of individual life and interaction remains Torsten Hägerstrand, "Space, Time, and Human Conditions," in A. Karlqvist (ed.), *Dynamic Allocation of Urban Space* (Farnborough: Saxon House, 1975); as well as the application of those ideas by Anthony Giddens, *The Constitution of Society* (Berkeley: University of California Press, 1984), 110–122.
6. Classical social theory is replete with this theme. See Georg Simmel, "The Metropolis and Mental Life," in *The Sociology of Georg Simmel* (New York: Free Press, 1957); and Louis Wirth, "Urbanism as a Way of Life," *American Journal of Sociology* 44 (July 1938).
7. For an excellent discussion, see Richard Sennett, *The Uses of Disorder: Personal Identity and City Life* (New York: Knopf, 1970), 137–188.

8. For an important discussion on this distinction, see James A. Christenson, "Gemeinschaft and Gesellschaft: Testing the Spatial and Communal Hypotheses," *Social Forces* 63 (1) (1984): 160–168.
9. For an important discussion, see David Harvey, *The Condition of Postmodernity: An Enquiry into the Origins of Cultural Change* (Oxford: Blackwell, 1990), 201–239.
10. Georg Simmel, "The Metropolis and Mental Life," in *The Sociology of Georg Simmel* New York: Free Press, 1957), 409.
11. See Richard Sennett, *The Fall of Public Man* (New York: Knopf, 1974) as well as Robert Putnam, *Bowling Alone: The Collapse and Revival of American Community* (New York: Simon & Schuster, 2000).
12. Georg Simmel, "The Metropolis and Mental Life" and Tony Hiss, *The Experience of Place* (New York: Random House, 1990).
13. Erving Goffman, *Frame Analysis: An Essay on the Organization of Experience* (Boston: Northwestern University Press, 1974), 22.
14. Michel de Certeau, *The Practice of Everyday Life* (Berkeley: University of California Press, 1984); David Harvey, *The Condition of Postmodernity* (Oxford: Blackwell, 1990); Tracy Strong and M. Hénaff, eds., *Public Space and Democracy* (Minneapolis, MN: University of Minnesota Press, 2001).
15. Goffman, *Frame Analysis: An Essay on the Organization of Experience*, 10.
16. Giddens, *The Constitution of Society*, 258–290; Harvey, *The Condition of Postmodernity*.
17. See Amy Kenyon, *Dreaming Suburbia: Detroit and the Production of Postwar Space and Culture* (Detroit, MI: Wayne State University Press, 2004) and Roger Salerno, "Alienated Communities: Between Aloneness and Connectedness," in L. Langman and D. Kalekin-Fishman (eds.), *The Evolution of Alienation* (Lanham, MD: Rowman and Littlefield, 2006), 253–268.
18. See Douglas Massey and Nancy Denton, *American Apartheid* (Cambridge, MA: Harvard University Press, 1998) and Iris Marion Young, "Segregation and Differentiated Citizenship," *Citizenship Studies* 3 (2) (1999): 237–252.
19. For an attempt at theorizing this aspect of suburbanism and democratic life, see Susan Bickford, "Constructing Inequality: City Spaces and the Architecture of Citizenship," *Political Theory* 28 (3) (2000): 355–376; Lizabeth Cohen, *A Consumer's Republic* and Margaret Kohn, *Brave New Neighborhoods: The Privatization of Public Space* (New York: Routledge, 2004).
20. See Carole Greenhouse, "Signs of Quality: Individualism and Hierarchy in American Culture," *American Ethnologist* 19 (2) (1992): 233–254.
21. For an interesting discussion of this with respect to gender, see Anne Markusen, "City, Spatial Structure, Women's Household Work, and National Urban Policy," *Signs* (Spring 1981): S23–S43; Melissa Gilbert, "'Race,' Space, and Power: The Survival Strategies of Working Poor Women," *Annals of the Association of American Geographers* 88 (4) (1998): 595–621; as well as Kraack Kenway and Hickey-Moody, *Masculinity Beyond the Metropolis* (New York: Palgrave Macmillan, 2006). With respect to race, see Loic Wacquant, "Urban Outcasts: Stigma and Division in the Black American Ghetto and the French Urban Periphery," *International Journal of Urban and Regional Research* 17 (3) (1993): 366–383; and Ali Madanipour, "Social Exclusion and Space," in A. Madanipour, G. Cars, and Judith Allen (eds.), *Social Exclusion in European Cities* (New York: Routledge Press, 2005), 75–94.
22. See Sennett, *Fall of Public Man*, as well as Eli Zaretsky, *Capitalism, the Family, and Personal Life* (New York: Harper and Row, 1976).
23. George Herbert Mead, *Mind, Self, and Society* (Chicago: University of Chicago Press, 1962), 134.
24. Ibid., 272.

25. For an important discussion of the role of the intersubjective mind's capacity to constitute the objective social world, see John Searle, *Mind, Language, and Society* (New York: Basic Books, 1998), 111–134 as well as Jürgen Habermas, *Justification and Application: Remarks on Discourse Ethics* (Cambridge, MA: MIT Press, 1994), 113–132. Although these thinkers do not provide a role for the effect of space upon intersubjective practices and the nature of mind, it is my explicit argument here that spatial organization is a crucial component of the ways in which intersubjectivity itself is structured.
26. The empirical evidence for this can be seen in voting statistics when broken down by region. See S. McKee and D. Shaw, "Suburban Voting in Presidential Elections," *Presidential Studies Quarterly* 33 (1) (2004): 125–144; Lang, R. and T. Sanchez, "The New Metro Politics: Interpreting Recent Presidential Elections Using a County-Based Regional Typology," *Metropolitan Institute at Virginia Tech Election Brief*, 2006: http://www.mi.vt.edu/uploads/NationalElectionReport.pdf (accessed: June 19, 2008); and J. Gainsborough, *Fenced Off: The Suburbanization of American Politics* (Washington, DC: Georgetown University Press, 2001).
27. See C. Weber and C. Federico, "Interpersonal Attachment and Patterns of Ideological Belief," *Political Psychology* 28 (4) (2007): 389–416.
28. Ibid.
29. See Stanley Feldman and Karen Stenner, "Perceived Threat and Authoritarianism," *Political Psychology* 18 (4) (1997): 741–770.
30. Stanley Feldman, "Enforcing Social Conformity: A Theory of Authoritarianism," *Political Psychology* 24 (1) (2003): 41–74.
31. This is also related to themes of anti-intellectualism. See Richard Hofstadter's discussion of the distinction between "intellect" and "intelligence" in his *Anti-Intellectualism in American Life* (New York: Vintage, 1963), 24–51.
32. This, of course, is the argument of Richard Sennett. See his excellent studies, *Families against the City* as well as his more developed discussion in *The Fall of Public Man* (New York: Knopf, 1977), 177–183. For a more contemporary discussion of the connection between nonurban life and conservative political ideas in American society, see Michael J. Thompson, "America's Conservative Landscape," in Michael J. Thompson (ed.), *Confronting the New Conservatism: The Rise of the Right in America* (New York: New York University Press, 2007).
33. There is a large literature on the ways that suburban space isolates individuals by means of structural design. See Kenneth Jackson's seminal study *Crabgrass Frontier: The Suburbanization of the United States* (New York: Oxford University Press, 1985); as well as Peter Katz, *The New Urbanism* (New York: McGraw-Hill, 1994); Philip Langdon, *A Better Place to Live: Reshaping the American Suburb* (Amherst, MA: University of Massachusetts Press, 1994); James Kunstler, *The Geography of Nowhere: The Rise and Decline of America's Man-Made Landscape* (New York: Simon & Schuster, 1993); Peter Calthorpe, *The Next American Metropolis* (New York: Princeton Architectural Press, 1993); and Andres Duany and Elizabeth Plater-Zyberk, *Towns and Town-Making Principles* (New York: Rizzoli, 1991).
34. Lewis Mumford, *The City in History* (New York: Harcourt, Brace, & World, 1961), 486.
35. Richard Sennett, *The Fall of Public Man: On the Social Psychology of Capitalism* (New York: Vintage, 1974) and *The Uses of Disorder* (New York: Vintage, 1970). Also see Philip Slater, *The Pursuit of Loneliness: American Culture at the Breaking Point* (Boston: Beacon Press, 1970).
36. Richard Sennett, *The Uses of Disorder* (New York: Knopf, 1970), 70.
37. For an excellent discussion of the role of fear in the rise of suburban culture, see Robert M. Fogelson, *Bourgeois Nightmares: Suburbia, 1870–1930* (New Haven: Yale University Press, 2005), 117–200.
38. See Karen Stenner, *The Authoritarian Dynamic* (New York: Cambridge University Press, 2005), 52–84; as well as Feldman, "Enforcing Social Conformity: A Theory of Authoritarianism."

39. For more on the lack of social conflict in suburban life, see M.P. Baumgartner, *The Moral Order of a Suburb* (New York: Oxford University Press, 1988).
40. Ibid., 72–100.
41. Douglas Massey and Nancy Denton, *American Apartheid* (Cambridge, MA: Harvard University Press, 1998).
42. For an excellent discussion, see J. Eric Oliver, *Democracy in Suburbia* (Princeton: Princeton University Press, 2001).
43. See Sennett's excellent book, *The Hidden Injuries of Class* for more on this aspect of work life (New York: Knopf, 1972).
44. This is an elaboration of Maurice Stein's insights in his book, *The Eclipse of Community* (New York: Harper and Row, 1960).
45. Louis Wirth, "The Scope and Problems of Community," in Albert Reiss (ed.), *Wirth on Cities and Social Life* (Chicago: University of Chicago Press, 1933), 166–177.
46. For an attempt at theorizing this aspect of suburbanism and democratic life, see Susan Bickford, "Constructing Inequality: City Spaces and the Architecture of Citizenship," *Political Theory* 28 (3) (June 2000): 355–376. Of course, there are also the economic theories of local public goods that point in this direction as well. See Charles Tiebout's classic paper, "A Pure Theory of Local Public Expenditures," *Journal of Political Economy* 64 (October 1956) as well as Joseph Stiglitz's critical extension of this model, "The Theory of Local Public Goods," in M. Feldstein (ed.), *The Economics of Public Services* (London: Macmillan, 1977).
47. Dolores Hayden, *Building Suburbia: Green Fields and Urban Growth, 1820–2000* (New York: Vintage, 2003), 128. Also see Cohen, *A Consumer's Republic.*
48. For an excellent collection of essays on this theme with respect to New York City, see Jerilou Hammett and Kingsley Hammett, eds., *The Suburbanization of New York* (New York: Princeton Architectural Press, 2007).

CHAPTER TWO

Antiurbanism in the United States, England, and China

ROBERT A. BEAUREGARD

From the late nineteenth century to the middle of the twentieth century, the countries of Europe and North America experienced unprecedented industrialization and urbanization. One consequence was to shift the center of cultural gravity away from the countryside, a shift often accompanied by antagonism toward the cities. Consequently, antiurbanism has been documented for the United States, England, and China, and its presence noted for Germany, Canada, Finland, Italy, Japan, and Russia. Yet, comparative studies that might reveal commonalities and differences across countries are few.[1]

The purpose of this chapter is to explore the material underpinnings of antiurbanism—a cultural and political sensibility derived from historically specific conditions, forged out of contentious relationships, and cast in an explicit spatial form.[2] To do so, I focus on three, emblematic instances of antiurbanism that occurred prior to and just after World War II; that is, before urban decline and shrinkage became relatively common. After that war, countries like the United States and England experienced a form of antiurbanism driven more by the deterioration of the large, industrial cities and the rise of the mass suburbs than any cultural clash between rural and urban interests.[3]

The three cases are the United States during the late nineteenth and early twentieth centuries when urban economic and political elites established dominance over rural interests, young adults left the countryside for the big cities, and the morality of life in the cities came under scrutiny; England during the same period when a realignment of class positions led to the nostalgic formulation of an "English way of life" set in the countryside; and China beginning in the late 1940s when Mao Zedong and the Communist Party took control of the national government and for political reasons championed rural life. Each of these examples highlights a different set of contributing factors. The United States's case represents a clash between rural and urban, political and economic interests expressed in cultural terms. By contrast, the English case focuses primarily on class relations and collective identity, while the Chinese case pivots on a program of national political and economic development.

Before proceeding to the cases, a few caveats are in order. Each points to a limitation of my approach and thus to future research possibilities. First, I am interested in antiurbanism as a discursive practice, not as a statistical generalization to be applied to citizen's cultural attitudes or as an explicit policy stance of governments. What is at issue is whether intellectuals and/or the popular media find antiurbanism to be a useful concept for framing or explaining changes in material conditions. Second, although I compare cases, no attempt is made to document the extent of antiurbanism across countries; these cases are not meant to be representative of all possible instances. Third, I leave aside the conundrum of why some highly urbanized countries (such as Australia) have no tradition of antiurbanism while others (such as the United States) do.

Fourth, antiurbanism almost always exists in a dialectic mode; that is, copresent with prourbanist attitudes. Antiurbanism draws its meaning from urbanism.[4] Neither the relative intensities of these conflicting attitudes nor their distribution across the population is addressed. The juxtaposition might well create ambivalence rather than stark and contradictory sensibilities. Finally, because numerous countries have had similar developmental experiences and intellectuals and ideas flow across national boundaries, antiurbanism in one country might well be related to antiurbanism in another. While I am intrigued by this possibility, it will have to await another investigation.

United States: Economics, Morality, and Politics

It is a rare commentator, one quite out-of-step with the prevailing sentiment, who avows that Americans have loved their cities. Rather,

Americans are commonly portrayed as lacking a "persistent or pervasive tradition of romantic attraction" to urban life.[5] The result is either a militant antiurbanism of hostility and neglect or a profound ambivalence. Whether pointing to the relentless westward expansion of the eighteenth and nineteenth centuries, the massive suburbanization of the twentieth century, or the enduring disdain of intellectuals, the general sense is that the cultural code on this matter was fixed at an early point in the country's history.

To the extent that it was pervasive, antiurbanism in the United States had roots in both a preindustrial, agrarian myth and the dislocations produced by the rise of cities as commercial centers, the vast rural-to-urban migration attendant to industrialization, and the political consequences of rising numbers of urban voters and legislators. Artists and intellectuals are a significant part of the story, but they are only one part. The spatial realignments of economic activity and political representation that centered antiurbanism were less a matter of intellectual complaint than a rearrangement of power relations. U.S. antiurbanism was about the wrenching shift from an economy anchored in agriculture and self-sufficient farmers to one based in manufacturing and the working class and industrial elites of the city. To this extent, it was a cultural formation most energized in the nineteenth century.

During the eighteenth century, an agrarian myth celebrated the moral superiority of the countryside.[6] A conceit mainly of intellectuals and the educated classes, this myth reflected the anxieties engendered by the commercialization of farming and the loss of an ostensibly innocent and independent way of life. In the nineteenth century, this agrarian myth became a "mass creed." As one commentator has noted, "the back-to-nature movement shifted from being a luxury of the rich to a preoccupation of an urban middle-class."[7] Rooted in pastoralism, it was supported by the belief that the countryside and the city embodied "diametrically opposed values" with the city posing a threat to rural life.[8] Romantic writers such as Ralph Waldo Emerson and Henry David Thoreau rejected the artifice of the city. They believed that by destroying solitude, the city stifled poetry, philosophy, and personal transcendence. Nature was the touchstone for a moral existence.

The growth of manufacturing and its spatial concentration in the cities prior to the Civil War (1861–1865) eventually subordinated agricultural activity to the demands of urban consumers and the manipulations of city-based financiers, corporate managers, and urban political interests.[9] Farming was commercial farming with all of the market forces attendant to such activity. The yeoman farmer—self-sufficient and "free"—was less and less common and the agrarian myth of an unmediated contact with nature and independence from societal constraints could no longer be dissolved in nostalgic yearnings. The farmer had become a businessman; a large proportion of his income was derived from land speculation as opposed to bringing crops to market. The objective conditions for the agrarian myth had been erased. By the Civil War, the countryside and its agriculture had become commercialized.[10] Antiurbanism had turned into an ideological weapon deployed by a retreating, agrarian society. The rise of a commercial middle class in the cities ruled over the class of "independent" farmers.

The end of the war and the further concentration of manufacturing in the cities combined with a new round of immigration to lead to further urbanization. As the cities prospered and agriculture mechanized, the cities became increasingly attractive to rural youth looking for economic opportunity. The exodus to the cities broke up rural families and was another sign that independent farming was at its end. Resentment on the part of those still clinging to rural life fed both a growing antiurbanism and a romantic rural nostalgia. The expanding incongruity between the moral status of the countryside and its economic position exacerbated those feelings.

These resentments and confusions were intensified by the physical and moral dangers of the city. There, seemingly innocent youth were tempted by alcohol and prostitution. Gambling, dance halls, and barrooms became a respite from the hard drudgery of factory labor. Low wages, unsafe working conditions, and unscrupulous employers and landlords made city life unpredictable and harsh and spread misery among recent arrivals. Churches could not expand fast enough or reach far enough to counteract the sin and deprivation.[11] The task at hand was thus to bring the moral order of the countryside to the dense, chaotic, and morally threatening cities. This responsibility was taken up by the new commercial and professional classes.[12]

The rising commercial class was less tolerant of alcohol, gambling, and the pleasures of the flesh than the patrician elites. During the Gilded Age of the late nineteenth century, the "evangelical creed and self-disciplined habits" of the middle class led to numerous social reform efforts, all designed to assert control over the city and its new inhabitants, former rural residents and immigrants.[13] These efforts had a strong moral dimension and included coercive activities such as the Temperance Movement and antiprostitution crusades as well as activities focused on improving the environment through housing reforms, parks and playgrounds, and civic beautification. Moral reform built on such earlier efforts as the Sunday School Movement and Bible Societies. Less explicitly religious, Progressive reformers were no less intent on changing the morals of those in harm's way. The city was "a menace to be subdued" and the white, Protestant, native, middle class was intent on doing so. Their targets were the slums and the lower and working-class African Americans and foreigners who occupied them. As one of their leaders proclaimed, the perils facing Christianity—immigration, intemperance, socialism, excessive wealth, and the Roman Catholic threat to the separation of church and state—were enhanced and "focalized" in the cities. More pointedly, "moral and religious influences are peculiarly weak at the point where our social explosives are gathered."[14] Antiurbanism took on a new character. Deeply ambivalent, apprehension about the city was tempered by a belief that reform was possible.

Intellectuals rued the deflation of the agrarian myth but were captivated by the shame of the cities.[15] The romantic writers of the decades before the Civil War were either in search of civilization or still clinging to the possibility of using Nature to resist the strictures of an increasingly interdependent society. Their anxiety drew from a sense that political insurrection in the cities was imminent, a position embraced by Thomas Jefferson in the eighteenth century. Jefferson feared the mob and contrasted it with the self-sufficient, yeoman farmer. Alexis de Tocqueville expressed similar views.[16] After the Civil War and through to the early years of the twentieth century, with industrialization robust and rural migrants and immigrants entering the cities in larger numbers, novelists still wrote of foreigners, socialism, and revolt, but they also condemned the city's inhumanity and called for reform. They juxtaposed the city as a place of promise and fulfillment with the reality of undeniable misery and shattered dreams. This disappointment was central to the cultural pessimism that came to haunt them.[17] More and more, the agrarian vision faded as an alternative. Reform took its place as the cultural attitude of choice.

By the 1920s, the moral superiority of the countryside had been abandoned. Widely recognized was that "the rise of the city was not necessarily a signal for the inevitable collapse of morality and social order."[18] Antiurban impulses such as Prohibition, the Ku Klux Klan, the Fundamentalist Movement, and the paintings of Norman Rockwell were seemingly balanced by countertendencies such as the writings of the sociologists Robert Park and Louis Wirth, the emergence of the urban detective novel, and the rising centrality of cities as places of mass entertainment such as dance halls, movie palaces, and night clubs.

During a decade of spreading affluence, a "celebratory tone toward the diversity and openness of urban life pervade[d] much of American social thought."[19]

This transition was accompanied by a major change in the locus of political authority in the federal and state governments. The growth of population in the cities relative to that in the countryside threatened rural control of state legislatures and the U.S. Congress. State legislators in particular acted to suppress urban interests by undermining their representation.[20] Legislative malapportionment, however, was a defensive maneuver destined to fail. The declining economic status of the countryside and the inability of agrarian romantics and others to assert their moral superiority over the cities was amplified by the redirection of government largess to manufacturing, city-based reforms, and urban infrastructure. Agriculture and rural areas were not neglected, but their voices were weaker than they had once been. When the U.S. Census announced in 1920 that the majority of the country's inhabitants lived in urban areas, the changeover from a rural to an urban society was official. And while this in no way prevented a version of antiurbanism from lingering, or even erupting again in the 1960s and 1970s, the period when antiurbanism drew its energy from the twinned moments of industrialization and urbanization had ended. With it, the agrarian myth was put to rest. Left behind was "the middle realm" that would be so frenetically reworked during the decades of postwar suburbanization.[21]

The case of antiurbanism in the United States thus constitutes a prime example of a cultural formation deeply embedded in "a passing phase of early urbanization and industrialization."[22] Grounded in that time, it was not, however, confined to it, for it had roots in a prior agrarian myth and that myth even reappeared, though weakened and transformed, in the decades after World War II. American "culture, as well as [its] literature, has tended continually to perceive environment through a pastoral filter."[23] The spatial restructuring attendant to the rise of manufacturing and corporate capitalism had social and moral as well as political consequences. It represented not just a realignment of the values attached to different spaces—the countryside, the frontier, the northeast region, the South, the cities—but the ascendance of a new set of elites. In its earliest manifestation, the self-sufficient farmer, large landowners, and intellectuals clinging to agrarian possibilities were dominant. By the end of this period, an urban middle class comprised of middle-class professionals and commercial and industrial elites had taken center stage.

Englishness and Rural Nostalgia

English antiurbanism exalted the countryside and left the city deep in the background, neither aggressively condemned nor forthrightly rejected. Rather than drawing its motivations from the consequences of industrialization and rapid urbanization, as was the case in the United States, it was based more in class differences, particularly those associated with the arts and popular leisure. In short, English antiurbanism was implicit rather than explicit and alternative rather than oppositional.[24] Its origins were in the cultural consequences of changing class relations.

Prior to World War II, English antiurbanism had two, distinct phases.[25] The first lasted from the 1870s to approximately 1914, the start of World War I. Then, a specific rural ideology was crafted around an ideal English life and an intense and negative reaction to the grime, chaos, and poverty of the industrial city. The second phase lasted from the end of World War I through to the start of World War II. It was centered on the countryside as a space of leisure and as a place of residence for professional, commercial middle class, and industrial working-class households. These rural cultural formations are widely—though not unanimously—accepted by observers of English national identity and are often loosely equated with antiurbanism.

By the eighteenth century, England had "a strong, generally affluent and increasingly integrated network of towns" and town life was accepted as part of the agricultural economy.[26] Subsequently, it was the first country in the world to experience rapid urbanization driven by the rise of manufacturing and the shift away from agriculture. The industrial cities of the late nineteenth century, however, sparked fears of a rebellious urban proletariat and deep concerns for the poverty, congestion, ill health, and dangers that came to be associated with places like Manchester, Liverpool, and London. As English goods-production began to decline in the late nineteenth century in the face of an expansion of U.S. manufacturing, the English economy shifted back to its commercial roots and was extended along the geographical trade routes opened up by British imperialism. As a further consequence of these changes, political and economic elites turned to the countryside as a source of cultural stability and economic opportunity only to discover that outmigration had severely diminished the able-bodied population and that the agricultural economy had atrophied. Two crises thus emerged, one urban and the other rural, and together they gave rise to nostalgia for the English village and the country life it represented.[27]

Rural nostalgia brought together gendered notions of domesticity, nationalistic yearnings for the quintessential English landscape, and a longing for a way of life that was less tainted by work and commerce and centered more on cultural pursuits. In contrast with the racial degeneracy of the cities where the "British stock" was being polluted by unwashed masses, the countryside was viewed as civilized. Unsurprisingly moral in tone, this rural nostalgia hardly mentioned the industrial city even though its denizens always lurked menacingly in the background as the incorrigible "other." Country people were the essence of England and morally superior to those working and living in the cities. The village with its shops and cottages, leisure pursuits such as hunting, and country squires who represented a yearned-for preindustrial rural hierarchy gave material shape to this antiurbanism.[28] Nature was an aesthetic object to be appreciated rather than a productive asset, an attitude that had much to do with the rise of the preservation movement and its goal of protecting the countryside from being despoiled and lost.[29]

No surprisingly, rural nostalgia had a decidedly regional bias. As memory, it was almost wholly anchored in the "south country" thereby excluding the regions north of London (where most of the major manufacturing cities were located) as well as Ireland, Wales, and Scotland.[30] From the architecture, landscape, history, and culture of this region were drawn the elements of a particular version of Englishness. In this way, class became a key dimension of this rural nostalgia. The "south country" was where aristocrats had the estates on which many rural workers found employment. Standing in contrast were the newly wealthy who had become prosperous from manufacturing, finance, and trade. Wishing to retain a romanticized version of an earlier way of life, the proponents of rural nostalgia were trying to imagine a bourgeoisie that could escape both the philistinism of capitalism and the depravity of the cities.

After World War I, the cultural embrace of the countryside spread beyond the upper middle class. The affluence of the 1920s and the development of council housing estates in the urban periphery combined to dampen rural nostalgia. Abandoning an aesthetic and class-based valuation of England's rural lands, the middle class and working class instead viewed the countryside as a place of leisure and a respite from the everyday drudgery of city life. The countryside had become an urban amenity. Fueled by national distress regarding the health of the country's youth (a concern related to the need for able-bodied soldiers in times of war), labor legislation that freed up time away from work, and general prosperity, more and more people began to engage in rural pursuits.[31] Rambling (including political rambling to challenge property rights), youth movements (e.g., the Boy Scouts), pilgrim walking, automobile excursions (spurred by the spread of car ownership), and weekend homes were all part of the redefinition of the countryside. Participants in these activities, moreover, were not the cultural elite of the middle class but a motley collection of modern dancers, hikers, health campaigners, fitness enthusiasts, nature mystics, and assorted hedonists. English antiurbanism ignored its urban "other." The early English suburbs and rural leisure pursuits were not rejections of urbanism, but rather extensions into what some scholars have labeled a "posturban" cultural formation.[32]

In the 1930s, with totalitarianism lurking in Europe, no more so than in Italy and Germany, order and stability were highly valued. The English way of life as a rural way of life provided the reassurance that those values would be protected. And, they were the government developed preservation laws and planning regulations to control development, and planning became one mechanism for achieving the security that the English desired. Emerging during a time of rural nostalgia and a cultural predisposition for the village and the cottage, planning took on a rural bias. Its major impact, though, occurred only after World War II, for even as the country was rebuilding its bombed-out cities, planners were pursuing a suburban middle ground at the urban fringe.[33]

Numerous scholars claim that this English antiurbanism has a long history and has survived unabated since at least the later nineteenth century.[34] Most of the literature also seems oblivious to parallel cultural formations that accept urbanism as a significant part of the English way of life, a condition that scholars have traced back to the thirteenth century.[35] The lone dissenting voice is that of Peter Mandler.[36] Mandler argues that English antiurbanism has been overblown and he blames the historian Martin Wiener. In *English Culture and the Decline of the Industrial Spirit, 1850–1980*, Wiener argued that in the late nineteenth and early twentieth centuries English culture, repulsed by the disruptions of the industrial revolution, spawned a middle class that devalued growth and innovation. Instead, that middle class patronized a slower, and rural way of life that nurtured a conservative attitude toward progress and legitimized antimodern sentiments. This attitude inhibited British economic development. Consequently, the country's economic troubles of the 1970s can be traced back to an antiurban, antimodern, antitechnology cultural disposition.[37]

Mandler disagrees. He accuses Wiener and other proponents of this "Englishness" thesis of presentism and of exaggerating the position of the rural way of life in the English cultural imagination. For Mandler, the attribution of Britain's economic woes during the 1970s and 1980s to a cultural formation of the late nineteenth century represents an "agenda of the present" that draws on the rise of conservatism, the pessimism engendered by the collapse of manufacturing and shipping in the industrial cities, and the weak state of the national economy.[38] In its place, he offers a more nuanced reading of the history of English antiurbanism that situates its pre–World War I phase in a small segment of the middle class, what he labels "tiny romantic minorities" and an "aesthetic auxiliary" to the dominant classes.[39] Writers, painters, poets, and other artistic types reacted against the aristocracy's disdain for the past, a disdain that threatened the traditional rural way of life and ignored the need for preservation. These artists and intellectuals were further dismayed by the accommodation to the city exhibited by professionals, merchants, financiers, and industrialists. This class segment of "cultural workers" constructed the rural nostalgia that eventually became associated with Englishness. Antiurbanism was thus a product of an intraclass conflict not a widely shared attribute of English culture. It owed its prominence, moreover, to the fact that it fit with the interests of another segment of the middle class, a growing financial and commercial elite. As manufacturing became less dominant, the English economy turned to its commercial roots; class segments whose wealth rested on trade and finance became more prominent. That these groups had ties to the aristocracy and wished to imitate its lifestyle turned them to the countryside where they could acquire country estates. Rural nostalgia provided cultural legitimacy.[40]

Prior to World War I, English antiurbanism was constructed out of the cultural perceptions of a specific class segment and was part of a class restructuring attendant to England's period of industrial city reform, its declining competitiveness in manufacturing, and its turn to imperialism and trade policy. Between the wars, the earlier rural nostalgia was picked up by an expanding middle class that used it to redefine the English way of life less as a matter of aesthetics and more as a matter of suburban living and leisure pursuits. Englishness was popularized. Lurking always beneath the surface of these cultural sensibilities—propping up rural nostalgia—was the English acceptance of the urban condition as "permanent and normal." Well before the countryside was romanticized, England had "come to terms with its urbanity."[41]

China: Counterrevolutionaries and Industrialization

The Chinese case continues the theme of national identity, though without the strong class component exhibited for England. Rather, from approximately 1949–1976, cultural nationalism was subordinated to a state-based political-economic project directed at modernizing the economy and protecting the state from counterrevolutionary forces. The Chinese case is also an instance—again, in contrast to the English case—of a problematic antiurbanism whose existence is only weakly supported by the evidence. Taking a well-defined and extreme position, one scholar claims that it is a conceit based on a "Western susceptibility to agrarian utopias and oriental fantasy," and factually wrong.[42] Others are skeptical, with one scholar mounting a positivist critique that contrasts antiurbanism as an interpretive scheme with a seemingly more credible argument about how Chinese state policy was driven by the need to achieve urban manageability and military preparedness.[43] These state imperatives, he claims, have been misread by some scholars intent on discovering an antiurban bias. To these dissents must be added China's "four thousand years [of] urban experience, probably longer on a continuous basis than that of any society."[44]

I want to set aside this debate for the moment and focus instead on the argument "for" Chinese antiurbanism. Rooted in that country's "ancient agrarian" tradition, antiurbanism was manifested in the Communist takeover of the late 1940s and its subsequent efforts to modernize the country.[45] Consequently, antiurban scholars draw on Mao Zedong's revolutionary connection to the peasantry, Communist Party fears of the counterrevolutionary inclinations of city-based intellectuals and functionaries, the tainted reputation of Chinese cities due to their status as treaty ports and thus foreign enclaves, and the Marxist commitment to abolishing the antagonism between the town and the countryside, with the last having a significant impact on national development policy.[46]

The antiurbanism position is rooted in the peasant origins of Mao and his comrades and in their revolutionary opposition to the Kuomintang (or nationalist party) government of the early 1920s. Most of the Communist Party leadership came from the countryside or from small towns and favored China's prevailing model of urbanization wherein the towns were administrative centers for the countryside. By contrast, the large cities, with the exception of Peking, were treaty ports. They had been ceded to foreigners in the mid-nineteenth century to serve as export platforms. Primarily trading centers, these cities also became centers for machine manufacturing and the sites of Western institutions such as banks. Moreover, they were never connected to the countryside, having failed to penetrate the predominately rural, internal markets, and were disconnected from the larger society. Chinese society being relatively insular, these enclaves were tainted; urban-based merchants and artisans were considered less virtuous than rural folk. "The foreign presence was almost exclusively urban," and in the popular imagination, "foreigners" and "large cities" came to be equated.[47]

Added to this was the fact that the Communists were more successful at organizing opposition in the countryside than in the cities. Their efforts in the cities were hindered by the greater strength of the Kuomintang Party there, the small number of industrial workers, the dominance of functionaries and intellectuals, and the foreign presence. Triggered by the 1927 coup of the Kuomintang in Shanghai, the Communists withdrew from the cities, ceding them to the opposition and focusing instead on organizing in rural areas. Mao would later need the large cities to industrialize and modernize the country, but conditions dictated that the revolution focus on the peasantry. Later, the Community Party ruled China from Peking, "the only major city that was not a treaty port."[48]

When the Communist Party took over the government in 1949 and declared the People's Republic of China, it continued to be leery of the cities. Mao was concerned that functionaries, intellectuals, youth, and foreigners were "potential breeders of bourgeois counterrevolution."[49] Functionaries would resist central state directives, intellectuals would mount uncontrollable critique, youth would challenge authority, and foreigners would exploit the country's resources. Concentrated in the cities, these groups made the cities into a source of political uncertainty. This counterrevolutionary potential was subsequently diluted by the Great Leap Forward of 1957 that industrialized rural areas and led to the formation of live work communes and the emergence of an urban proletariat. In addition, the Cultural Revolution of 1966–1968 required urban elites and Communist Party cadres to relocate to the countryside to "serve the people." City elites would work the land and be educated by the peasants in May 7 Cadre Schools, thus developing the appropriate revolutionary values of egalitarianism and a zeal for combating revisionist tendencies. The "sending down" (*hsia-fang*) of urban residents thereby blocked the ability to organize against the regime and, combined with *hukou* registration that tied an individual's livelihood to a specific place, enabled the Communist Party to maintain managerial control over the cities. In addition, by moving labor, particularly educated youth, from the cities to the countryside, the supply of and demand for educated labor in the cities was ostensibly balanced at a time when China was still at an early stage of modernization.[50]

While such actions indicate a relatively strong antiurbanism, that conclusion has to be tempered by the Marxist commitment to removing the contradictions between the countryside and the city as well as the necessary role that cities played in modernization, particularly given the emphasis on industrialization. Karl Marx argued in the "Communist Manifesto" that "the bourgeoisie subjected the country to the rule of the town" and noted in *The German Ideology* that "the abolition of the antagonism between town and country is one of the first conditions of communal life."[51] Society could evolve to socialism only by eliminating the economic, political, and cultural exploitation of rural folk by city-based elites. Drawing much of his revolutionary ideas from Marx and Engels, Mao recast these ideas in terms of the "three major differences" or contradictions: urban and rural, mental and manual labor, and worker and peasant.[52] Cities, especially the treaty ports, would not be eliminated but rather integrated into the new Chinese society. The goal was to balance development between the coastal areas, where these cities were located, and the interior and thereby to consolidate the country's borders. Moving strategic investments to the interior also protected them from the devastation of a possible invasion by the Soviet Union, Korea, or Taiwan where the Kuomintang had fled. Consequently, the slogan for the early period of development policy alluded to, but did not clearly reflect, the traditional role of Chinese cities: "agriculture the base, industry the leading sector."[53] The cities would be contained so that industrialization could proceed.

From the beginning, Mao's approach to modernization rested on large-scale industrialization. And since industrialization would proceed faster with large factories and large factories would require geographically concentrated pools of labor, Mao could not wholly dismiss the big cities. However, the then-current makeup of the treaty ports would have to change from an emphasis on consumption to an emphasis on production. "Produce more, consume less" was the slogan, and the distinction was a veiled allusion to the decadence of foreigners and intellectuals.

In his quest to industrialize nationally, Mao also needed investment capital and this was another reason to contain consumption. Consumption would deflect resources from investment in industrial plant and equipment. Consequently, the amount of investment in nonproductive activities was limited and every effort was mounted to recycle materials and apply them efficiently. In fact, the latter was the primary goal of neighborhood workshops that found alternative uses for waste products from manufacturing, employed female labor, and contributed to the organization of neighborhood communes.[54] Minimizing infrastructure investment in the cities further fostered the sense that official Chinese policy was antiurban.

The First Five Year Plan (1953–1957), for example, aimed to decentralize industry from the coastal cities through new investment and the relocation of existing factories from urban centers. The interior would become less of an agricultural region. This would overcome the stark uneven development between the coastal cities and the rural interior. In support of decentralization were various policies (e.g., individual registration and ration certificates) to control rural-to-urban migration and to transfer city residents to the countryside. All of this enabled Mao to avoid the "giantism" of large cities and their demand for investment in heavy infrastructure. "Giantism," of course, would unbalance the development landscape.

Reading these events from a prourban perspective leads easily to accusations of antiurbanism. And, to the extent that the Communist Party embedded its policy in ideological arguments, often using slogans to garner peoples' attention and motivate them, attitudes toward the countryside and city became part of (state-controlled) cultural sensibilities. Yet, China is also a case that can support a contrary reading. Mao Zedong was not as much antiurban as looking for a way to even out development in order to combat the backwardness of the China that he had wrested from the Kuomintang. And, quite importantly, he needed to manage the political potential of the cities, a concern that Thomas Jefferson had expressed back in 1784 and that Mao saw as involving not "the mobs of great cities" but intellectuals and functionaries.[55]

Conclusion

Although these three cases share many similarities—industrial urbanization being the most obvious, they are not quite the same cultural formation. Antiurbanism in the United States was embedded in wrenching social transformations generated by the geographical reorganization of the country's political economy. In England, class differences rose to the surface as intellectuals and popular commentary exercised their right to define the core of the country's national identity. By contrast, politics dominate the Chinese case. Leary of intellectuals and functionaries in the cities, needing to balance development to unify the nation, and obsessed with Communist Party control, Mao Zedong acted in ways that suggest an antiurban bias. Chinese antiurbanism, moreover, was not rural nostalgia and has seemingly no popular extensions. To this extent, it does not fully qualify as a cultural movement.

The main point of this chapter is a simple one. Antiurbanism is more than one thing. And, when manifested in specific times and place, it reveals its diversity. The differences, moreover, help us to understand the place of urbanism in the national imagination. How scholars and intellectuals weave cities into cultural formations speaks to the historical and geographical particularities of nations. Such sensibilities and attitudes also shape how people and states act in an urban world.

Notes

1. For exceptions, see Andrew Lees, *Cities Perceived: Society in European and American Thought 1820–1940* (New York: Columbia University Press, 1985) and Carl E. Schorske, "The Idea of the City in European Thought: Voltaire to Spengler," in Oscar Handlin and John Burchard (eds.), *The Historian and the City* (Cambridge, MA: MIT Press, 1963), 95–113.
2. My premise is Leo Marx's observation that the city has become "an abstract receptacle for displaced feelings about other things." See Leo Marx, "The Puzzle of Anti-Urbanism in Classic American Literature," in *The Pilot and the Passenger* (Oxford: Oxford University Press, 1988), 210. In referring to cultural sensibilities and formations, I have in mind "meanings and values as they are actively lived and felt," what Raymond Williams labeled "structures of feeling." See his *Marxism and Literature* (Oxford: Oxford University Press, 1977), 132.
3. See Robert A. Beauregard, *When America Became Suburban* (Minneapolis, MN: University of Minnesota Press, 2006), 76–78 for the United States and Claire Colomb, "Unpacking New Labour's 'Urban Renaissance' Agenda: Towards a Socially Sustainable Reurbanization of British Cities?" *Planning, Practice and Research* 22 (1) (2007): 1–24 for England.
4. On such relationships, see Andrew Sayer, *Method in Social Science* (London: Hutchinson, 1984), 82–87. Raymond Williams takes this approach in his well-known *The Country and the City* (Oxford: Oxford University Press, 1973).
5. Morton White and Lucia White, *The Intellectual versus the City* (New York: New American Library, 1926), 13.
6. Richard Hofstadter, *The Age of Reform* (New York: Vintage Books, 1955), 23–31.
7. Peter J. Schmitt, *Back to Nature: The Arcadian Myth in Urban America* (Baltimore: Johns Hopkins University Press, 1969), 4.
8. James L. Machor, *Pastoral Cities: Urban Ideals and the Symbolic Landscape of America* (Madison, WI: University of Wisconsin Press, 1987), 5.
9. William Cronon, *Nature's Metropolis: Chicago and the Great West* (New York: W.W. Norton, 1991).
10. This argument comes from Hofstadter, *The Age of Reform*, 23–59.
11. Josiah Strong, *Our Country: Its Possible Future and Its Present Crisis* (Cambridge, MA: Belknap Press, 1963 [1886]), 171–186.

12. See Paul Boyer, *Urban Masses and Moral Order in America, 1820–1920* (Cambridge, MA: Harvard University Press, 1978) and Daphne Spain, *How Women Saved the City* (Minneapolis, MN: University of Minnesota Press, 2004), 63–86.
13. This discussion of moral order is based on Boyer, *Urban Masses and Moral Order in America, 1820–1920.* The quoted phrase is on p. 77.
14. Strong, *Our Country: Its Possible Future and Its Present Crisis*, p. 179.
15. The key reference here is White and White, *The Intellectual versus the City.*
16. Ibid., p. 34.
17. James W. Tuttleton, "City Literature: States of Mind," *Modern Age* 33 (3) (1990), 269–279.
18. Boyer, *Urban Masses and Moral Order in America, 1820–1920*, 287.
19. Ibid., 288.
20. Dennis R. Judd and Todd Swanstrom, *City Politics: Private Power and Public Policy* (New York: HarperCollins, 1994), 50–52.
21. Machor, *Pastoral Cities: Urban Ideals and the Symbolic Landscape of America*, 14. For post–World War II antiurbanism, see Robert A. Beauregard, *Voices of Decline: The Postwar Fate of U.S. Cities* (New York: Routledge, 2003), 150–178.
22. Joseph Harry, "American Anti-Urbanism and Its Evolution," *Journal of Voluntary Action Research* 7 (3–4) (1978): 36–43.
23. Machor, *Pastoral Cities: Urban Ideals and the Symbolic Landscape of America,* 4.
24. On alternative and oppositional cultures, see Raymond Williams, *Problems in Materialism and Culture* (London: Verso, 1980), 40.
25. This periodization follows that in Alun Howkins, "The Discovery of Rural England," in Robert Coles and Philip Dodd (eds.), *Englishness: Politics and Culture 1880–1920* (London: Croom Helm, 1986), 62–88.
26. D.M. Palliser, "Preface by the General Editor," in D.M. Palliser (ed.), *The Cambridge Urban History of Britain,* Vol. 1 (Cambridge: Cambridge University Press, 2000), xix–xxiii, quotation on page xxi.
27. Jan Marsh, *Back to the Land: The Pastoral Impulse in England, From 1880 to 1914* (London: Quartet Books, 1982).
28. Darrin Bayless, "Revisiting the Cottage Council Estates: England, 1919–39," *Planning Perspectives* 16 (2001): 171.
29. A.D. King, "Historical Patterns of Reaction to Urbanism: The Case of Britain 1880–1939," *International Journal of Urban and Regional Research* 4 (4) (1980): 453–469.
30. Howkins, "The Discovery of Rural England.".
31. A good description of rural leisure activities in this period can be found in David Matless, "'The Art of Right Living': Landscape and Citizenship, 1918–1939," in Steve Pile and Nigel Thrift (eds.), *Mapping the Subject: Geographies of Cultural Transformation* (London: Routledge, 1995), 93–122. See also Bayless, "Revisiting the Cottage Council Estates: England, 1919–39," 170–171; John Lowerson, "Battles for the Countryside," in Frank Gloversmoth (ed.), *Class, Culture and Social Change: A New View of the 1930s* (Sussex: The Harvester Press, 1980), 258–280; and Philip Lowe, Jonathan Murdoch, and Graham Cox, "A Civilized Retreat? Anti-Urbanism, Rurality and the Making of an Anglo-Centric Culture," in Patricia Healey, Stuart Cameron, Simin Davoudi, Stephen Graham, and Ali Madani-Pour (eds.), *Managing Cities: The New Urban Context* (New York: John Wiley & Sons, 1995), 63–82.
32. Peter Mandler, "Against 'Englishness': English Culture and the Limits to Rural Nostalgia, 1850–1940," *Transactions of the Royal Historical Society*, Sixth Series, 7 (1977): 171.
33. Bayless, "Revisiting the Cottage Council Estates: England, 1919–39" and King, "Historical Patterns of Reaction to Urbanism: The Case of Britain 1880–1939," on page 459 notes the national government's belief in the basic superiority of "one family, one dwelling."
34. See Bayless, "Revisiting the Cottage Council Estates: England, 1919–39"; Ruth Glass, *Cliches of Urban Doom* (Oxford: Basil Blackwell, 1989), 41–42; King, "Historical Patterns

of Reaction to Urbanism: The Case of Britain 1880–1939"; Lowe, Murdoch, and Cox, "A Civilized Retreat? Anti-Urbanism, Rurality and the Making of an Anglo-Centric Culture," 65; and Williams, *The Country and the City*, 2.

35. Palliser, "Introduction," 4, in *The Cambridge Urban History of Britain*.
36. Mandler, "Against 'Englishness': English Culture and the Limits to Rural Nostalgia, 1850–1940."
37. Martin J. Wiener, *English Culture and the Decline of the Industrial Spirit, 1850–1980* (Cambridge: Cambridge University Press, 1981).
38. Mandler, "Against 'Englishness': English Culture and the Limits to Rural Nostalgia, 1850–1940," 175.
39. Ibid., p. 168 and p. 166 respectively.
40. Lowe, Murdoch, and Cox, "A Civilized Retreat? Anti-Urbanism, Rurality and the Making of an Anglo-Centric Culture," 69.
41. The quoted phrases in this and the previous sentence are from Mandler, "Against 'Englishness': English Culture and the Limits to Rural Nostalgia, 1850–1940," 160.
42. R.J.R. Kirkby, *Urbanization in China: Town and Country in a Developing Economy, 1949–2000* (New York: Columbia University Press, 1985), 18. In contrast, see Laurence J.C. Ma, "Anti-Urbanism in China," *Proceedings of the Annals of the Association of American Geographers* 7–8 (1975–1976): 114–118.
43. George Chu-sheng Lin, "China's Industrialization with Controlled Urbanization: Anti-Urbanism or Urban-Biased?" *Issues & Studies* 34 (6) (1998): 101. See also Marina Basso Farina, "Urbanization, Deurbanization and Class Struggle in China 1949–1979," *International Journal of Urban and Regional Research* 4 (4) (1980): 485–502 and John Wilson Lewis, "Introduction: Order and Modernization in the Chinese City," in J.W. Lewis (ed.), *The City in Communist China* (Stanford, CA: Stanford University Press, 1971), 1–26.
44. Rhoads Murphey, *The Fading of the Maoist Vision: City and Country in China's Development* (New York: Methuen, 1980), 18. See also George C.S. Lin, "Chinese Urbanism in Question: State, Society, and the Reproduction of Urban Spaces," *Urban Geography* 28 (1) (2007): 12–14.
45. Kirkby, *Urbanization in China: Town and Country in a Developing Economy, 1949–2000,* 4.
46. The association of antiurbanism with the nationalist projects of authoritarian regimes has also been noted for Italy during the fascist period (1924–1945) under Benito Mussolini and in Germany during the Nazi era (1934–1945) under Adolf Hitler. See, respectively, Anna Treves, "The Anti-Urban Policy of Fascism and a Century of Resistance to Industrial Urbanization in Italy," *International Journal of Urban and Regional Research* 4 (4) (1980): 470–484 and Schorske, "The Idea of the City in European Thought: Voltaire to Spengler," 113.
47. Rhoads Murphey, "The Treaty Ports and Chinese Modernization," in Mark Elvin and G. William Skinner (eds.), *The Chinese City Between Two Worlds* (Stanford, CA: Stanford University Press, 1974), 67 [17–71].
48. Ibid., 71. Ironically, the Communist Party was founded in 1921 in Shanghai, the most notorious foreign port.
49. Rhoads Murphey, "Chinese Urbanization Under Mao," in Brian J.L. Berry (ed.), *Urbanization and Counter-Urbanization* (Beverly Hills, CA: Sage, 1976), 313.
50. Murphey claims in *The Fading of the Maoist Vision* that "China's use of *hsia-fang* is of course a reminder of its still basically anti-urban bias" (100). See also Ma, "Anti-Urbanism in China," 115–116. On the May 7 Cadre Schools, see Christopher L. Salter, "Chinese Experiments in Urban Space: The Quest for an Agrapolitan China," *Habitat, An International Journal* 1 (1) (1976): 19–35.
51. Karl Marx and Friedrich Engels, "Manifesto of the Community Party," in Lewis S. Feuer (ed.), *Marx & Engels: Basic Writings on Politics and Philosophy* (New York: Anchor

Books, 1959), 11 and Karl Marx and Friedrich Engels, *The German Ideology* (New York: International Publishers, 1970), 69.

52. Ma, "Anti-Urbanism in China," 115.
53. Kirkby, *Urbanization in China: Town and Country in a Developing Economy, 1949–2000*, 5.
54. Slater, "Chinese Experiments in Urban Space: The Quest for an Agrapolitan China," 28–32.
55. On the Jefferson reference, see White and White, *The Intellectual versus the City*, 25.

CHAPTER THREE

*The Origins of Antiurbanism**

JAMES A. CLAPP

Students of the so-called urban crisis have often made the observation that the urban problem is related more to problems in the city rather than the problem of the city. While the author would tend to agree with this statement within the context in which it was made, one may still wonder whether there exists, in real or imaginary terms, a problem of the city. Perhaps, there is not so great a dichotomy between the two perspectives, since our contemporary problems of the city may be or seem all the more intractable and difficult to deal with as a result of sentiments and predispositions that many of us hold about cities and urban life in general. If such is the case, the problems in the city, real as they may be, may function as a reification of negative images about city life in general, forcing many to see the urban way of life as less perfectible than it may be and to opt for departure at the earliest opportunity.

Antiurbanism is hardly a new subject for discussion, and my purpose here is not to conduct a thoroughgoing review of its abundant literature. Several good analyses already exist. However, some brief comments may assist readers who are unfamiliar with it. Philosophers, scientists, ecclesiastics, and literati of every age have expressed views on the cities of their times and the prospects for the future. Social thought in general is suffused with direct and indirect commentary on cities and city life. A wide variety of works have documented antiurban themes among the influential writers in ancient times, such as Plato and Aristotle, whose ideal cities were quite small in size, and agriculturally dominated because of the importance placed upon the farming classes.

Likewise, Roman-period writers stressed the essential importance and virtues of the farming class, emphasizing, as did many others, that the best and most reliable soldiers came from the agricultural sector.[1] Augustine's dichotomous view stands out, posing the city of man against the city of God. Later, influential thinkers and writers who either stressed distaste for the city or the primacy of rural and pastoral ways of life including Thomas More, Machiavelli, and Rousseau.[2] There were exceptions, of course, among them Aquinas, who felt that the city was the natural state of man, and Voltaire, who celebrated London in verse as the rival of Athens, and also before the nineteenth century, Adam Smith.[3] The urbanizing effects of nineteenth-century industrialism upon cities are well documented and spurred increased philosophical and literary commentary stressing antiurban themes.

The reader is urged to consult such works by the Whites, on intellectuals' negative attitudes toward cities, Leo Marx's discussion of antiurban themes in literature and Raymond Williams' examples of poetry.[4] Mass media holds many examples as well. Let the reader compare the highly popular television series, The Waltons, set in rural area during the Depression—a program that celebrates the simple life guided by the good book, the extended family, and honest labor in the soil—with the dominant programming in urban setting, which are principally crime and detective pulp. Indeed, if television programming has any basis in fact, the urban exodus might prove calamitous in spite of our National Defense Highway System.

Varieties of Antiurbanism

Antiurbanism is perhaps too general a term to convey the variety of negative urban imagery and opinion. For example, some of the ancient writers appear to stress the importance of the pastoral and agricultural classes more so than their negative opinions of the city, perhaps because of the proportion of these classes, which were needed to support even very small urban populations. Greek and Roman commentators appear to fall into this category. Others, such as Rousseau, flatly asserted the superiority of atavistic primitivism. Some, such as Mumford, bemoan the effects of urban technology, while Jefferson worries over problems of governance of large urban populations. Still others simply drew their dislike for the city from their distaste for aspects of it—its commerce, and boosterism, its insensate industry, its un-urbanized masses, and its lack of beauty.[5]

To some degree a distinction may be drawn between two general forms of antiurbanism in America. The first might be called Romantic Dis-Urbanism, as evidenced in the views of Thoreau, Hawthorne, Emerson, and Melville. Generally this view celebrates the natural over the man-made, and the rural and the wilderness over the city. Somewhat later, principally after the Civil War, antiurban sentiment tended to be based more upon the failure of American cities to live up to the ideals that a number of social thinkers compared urban conditions against. In this view, it is the distrust and scorn for commercialism, swelling urban masses, and other attributes of city growth that motivate the critiques of Henry Adams, Dreiser, Henry James, and John Dewey. Ironically, as White pointed out, the romanticists felt that the city was too civilized; the others found it not civilized enough.[6]

A second distinction might be drawn between those commentators whose concerns and negative sentiments are more directed to the process of urbanization and its relationship to the human condition as opposed to those who tend to place emphasis upon the burgeoning size of urban areas or the effects of city development upon environmental conditions.

Thirdly, there is a lack of cross-cultural analysis of general attitudes toward cities and urbanization. While other concerns about this subject have not allowed pursuit of a cross-cultural perspective, there are some indications that there may be more positive attitudes toward cities in Eastern societies. In particular, Wheatley has written extensively on the symbolic and metaphysical functions of cities in Asian societies.[7]

Opinion Polls

A second form of documentation of antiurban sentiments is opinion polls regarding location preferences. The preference for not living in cities, which is one index of these sentiments and attitudes, is rather consistently reflected in surveys of housing, neighborhood, and general locational preferences. For example, a survey reported by *Time Magazine* several years ago states: One of the most intriguing findings of all in the "State of the Nation" is that the majority of Americans yearn to escape urban areas not for suburbia, but for the truly open spaces. While only one out of three Americans now lives in towns, villages, or rural areas, more than half of the poll sample said they would prefer such a setting. That figure is swelled by the ranks of black city dwellers, seventy percent, that want to move out. Conclude the editors: "The figures suggest that if American people could follow their inclinations, the population of our cities would be cut in half. The proportion of suburbanites would remain the same. The proportion enjoying country life would more than double, from less that two in ten to almost four in ten."[8]

A 1968 Gallup poll indicated that only 18 percent of the respondents expressed a preference for living in central cities; 25 percent favor the suburbs; 29 percent opt for life in small towns, and, "...27 percent actually said they wished they could live on a farm."[9] As far as future housing consumers are concerned, surveys of American youth reflect these strong preferences for exurban living. A recent poll of students and nonstudents rendered the following breakdown of locational preferences for the question "In which type of place would you most like to live?"[10]

	Student	*Nonstudent*
Large city	11.4	14.2
Suburb	24.5	17.8
Small city or town	34.8	39.2
Rural area	26.1	28.5
No answer	2.2	0.3

While some surveys indicate that exurban tendencies may not be as strong among lower socioeconomic groups and inner-city racial and ethnic groups, these groups generally appear to share the housing and locational preferences of the American population at large. A small, but intensive survey of Black youth in Boston indicated that the overwhelming majority of the children surveyed preferred home characteristics of the country or suburbs when asked where they would like to live their adult lives. Fifty-four of the sixty participants wanted suburban housing: "a one-family house... with a big fence around it... a garden and a place where kids can play;" "a one-family house out in the suburbs with a big back yard and not too many neighbors around;" "A pretty fair size white house."[11] Lastly, results of a survey conducted on youth preferences in Sweden demonstrate that the desire to live outside cities is not restricted to Americans. Although Sweden is frequently cited by planners as a nation of well-planned cities and suburbs, its youth are apparently unimpressed. The Swedish Institute for Public Opinion Research reported that 83 percent of its respondents indicated the desire to live outside of cities. Half of this number preferred small towns; the other half would opt for the countryside. Only 14 percent expressed a desire to live in a large city.[12]

There are, of course, some good reasons to be cautious about generalizing to isolate on preference variables, such as living location, without dealing with the question of trade-offs. If responses are left unaltered by an exurban living location—accessibility to employment opportunities being the most obvious—the degree of strength of such preference cannot be adduced. Certainly the fact that many people do not appear to act upon these expressed preferences would appear to indicate that such trade-offs are potentially significant in explaining the differences between needs and wants and images and realities. Still the residue of a paradox remains: one of continued urbanization—although at technologically permitted low densities—and the persistence of varieties of ambivalent and negative feelings toward cities and urban life. The following discussion raises several hypotheses that speculate upon the origins and persistence of antiurban sentiments. The thematic varieties of antiurbanism offered here depart from the more typical views documented through opinion polls and philosophical thought; rather, they attempt to explicate alternative (but somewhat interrelated) causes for urban discontent by examining potential explanations that transcend locational preferences and focus more directly upon the existential consequences of the urbanization process.

Biblical Admonitions

Probably no theme is more influential in the Judeo-Christian view of the city that the Augustinian notion of the City of God versus the City of Man. Cities achieved a poor reputation early in biblical history by virtue of their association with Cain, who, after murdering his brother, forsakes the protection of God for the protection of the city. However, the association with homicide is only the beginning of a representation throughout the Old Testament of cities and city life as somehow being at odds with divine purposes.[13]

The history of the Hebrew people in cities is almost always one with a consequence of the erosion of faith or other stresses upon their covenant with Yahweh through contact with the pantheisms and pagan rites of urbanites. Jericho, Sodom, Gomorrah, and other Bilical cities represent failures of urban man to follow divine instruction, with perhaps the most representation of the inability of man to come closer to his god through cities in absurdity of the Tower of Babel. The point was fairly clear: man could only reach God through faith, not through his rational powers or technical capabilities. While other examples might be drawn, they would only belabor the point that the Old Testament largely depicts the city as an unnatural setting for man and one in which he places himself in defiance of God's will. The city became a symbol of the fall from grace, a notion that has functioned as a recurrent literary theme from ancient times to the present.

In tracing these same points Hadden and Barton add that these religious origins of antiurbanism are not restricted to the Judeo-Christian traditions. As evidence they cite the fourteenth-century Muslim thinker Khaldun whose analysis of the difference between nomadic and urban people stressed that, while the urban way of life led to high achievements in human development, urban populaces inevitably degenerated into corruption, self-indulgence, sexual perversions, and the loss of community and personal identity. The nomadic way of life was contrasted favorably with these conditions.[14]

Likewise, the antiurban posture of the Old Testament appears to derive largely from the fact that the chosen nation and its God were primarily nomads. The first antiurbanism may have had elements of antiagrarianism as well, since nomadism was the dominant way of life for thousands of years and the transition to permanent settlements, although for some time agrarian in form, represented a considerable break with social and theological traditions. The conflict between the nonplace and place-related gods has been noted in both theological and popular writings.[15] Residents of early settlements began to associate their deities with specific places in contrast to a universal God, and in part, cities began to focus the debate between monotheism and pantheism. Since the Bible is allegedly a work of divine inspiration, antiurbanism appears to originate in pretty high places. However, it is also possible to take the secular view that, since the Bible was the work of early religious leaders and prophets, the antiurbanism expressed in its pages represents recognition on their parts of the relationship between the secularization of urban man and the erosion of ecclesiastical authority. Cities meant the development of other powerful institutions and authorities, notably those of commerce and civil law.

However, the major threat to ecclesiastical authority and its interpretation of divine will appears to be that urban man gradually began to effect a view of the world that differed considerably from the ecclesiastical view. The ecclesiastical cosmology focused upon the temporariness of terrestrial life and stressed its preparatory role. Earth life was to be endured as much as lived. The tone of this interpretation was fatalistic; dominant mindset was resignation to the consequences of events. Tribal man was not supposed to ask many questions, as evidenced in the adage, "the Lord works his wonders in mysterious ways." This was an interpretation of events, along with the ministrations of religious leadership, which made even the most Jobian of lives seem purposeful and thereby tolerable.

Urban man is drawn to a different view. His view of events, while it may become deterministic, is far less fatalistic. He is not content to let things happen to him; rather, urban man believes he can cause events to happen in a purposeful way. In a real sense, then, urban man is Man the Planner, who attempts to take control of his destiny. The efficacy of his rational powers reinforces the notion that he has an active role to play in his nature, its modification, and its definition of purpose. Where it is retained his theological justification for this attitude is no longer based on a resigned acceptance of "mysterious wonders" of the world, but to the theological rationalism that God will reveal Himself to man through man's intellect. (A less noble thought perhaps more significant linkage is retained through the adage that "the Lord helps those who help themselves.")

With or without theological justification, this shift in cosmology is significant; man defines his needs and seeks to achieve them. This sense of open-endedness of being is perhaps most dramatically illustrated in the ways that urban man no longer considers his own physical and psychic nature as "given." Hearts are transplanted, sexes are "changed," conception prevented at will, consciousness and emotions pharmacologically altered. Chemists and biologists almost fearlessly pursue the secret of life itself and proponents of cryonic preservation and cyborgs challenge its finiteness. But while these means are considerable, they do not imply the correctness of ends; they only expand the range.

This perspective suggests not only that early antiurban sentiments may have been rooted in metaphysical debates, but also that negative attitudes toward cities may continue to derive from deeper human concerns than simply the stylistic preferences for pastoral landscapes and quaint farms and villages. To some degree urban man's curious attitudes toward the cities that he continues to build and inhabit may derive in some small measure from a residue of guilt over having ignored biblical warnings to avoid cities. Furthermore, it may well be that the scriptural view of the world and its universe offers plausible final causes, or sense of purpose and meaning, the need for which resides deeply imbedded in man's psyche. In coming to cities he has to some extent cut himself off from that view.

Existential Dilemmas

While the original antiurban perspective may have been authored by religious nonurbanities, it may have been based on sound intuitions (setting aside the view of divine inspiration) about the importance of meaning in the actions of men. Man's increasing knowledge about his own nature and the universe does not appear to have rendered any greater sense of the meaning of his existence. Despite the probing into his own molecular biology or into outer space, the search has led as much to existentialist meaninglessness as anything else. Cox provides another angle on this point: "Both tribal man and secular man see the world from a particular, socially and historically conditioned point of view. But modern man knows it, the tribal man did not, therein lies the crucial difference."[16] This realization "relativizes" the consciousness of urban-secular man. He recognizes that his values, his beliefs, his customs, and his thought patterns are a product of his environment and his history. He comes to understand that had he existed under different times and different conditions, he might have seen things differently. This process therefore appears to drift urban-secular man toward a relativistic view of the world. The result of this process is not only to subjectivize worldviews and values, but to "ephemerize" them as well. What is or what ought to be becomes no longer a matter of timeless objective truths, but a matter of time and circumstance. Historically this results in what might be another clue to the roots of antiurbanism. Urban-secular man recognizes the concomitant increased deprivation of continuity and cyclicality in urban life. The rate of attitudinal and social change deprives parents from passing on cherished values and skills to their offspring. The senior generation laments that values have changed in their lifetimes and that former taboos have become socially sanctioned if not recommended forms of behavior. In short, urbanization and secularization are processes that by their nature seem to create challenges and alterations in the preexisting order. They are processes that conceptually appear more linear than cyclical.

To conclude these points, it has been offered that the secularization of urban man results in a cosmology that not only generates alternative and competing views of reality, but also challenges any interpretation of events that assigns them meaning or purpose as being more than subjective or aesthetic judgment. This both causes and becomes reinforced by rapid social change and the relativization of values. This discussion therefore arrives at the implication that antiurbanism, or its counterparts in pastoral or agrarian romanticism, probably arise and persist because they offer in their theological and naturalistic imagery, final cause that impute meaning and continuity to life. They permit some derivation of universal principles and an ethical certainty. If such is the case, antiurbanism is something far deeper and complex than the desire for a single-family cape cod on a half acre. It represents a quest for an ordered and coherent cosmology and a sense of purposeful design.

The Psychobiology of Urban Life

A third and increasing popular hypothesis is that man's rate of urbanization may simply have outstripped his capability for psychological and biological adaptation. Several ethologists have reminded us in recent years that most of human existence has been spent outside of cities, or for that matter outside of permanent settlements. Some, such as Morris in his view of cities as "human zoos," stress that the psychological and genetic legacy of millennia are severely strained by the complex organization, competitiveness, and high compression of living of modern cities.[17]

Much of the end study of "behavioral sinks" is open to considerable dispute on several grounds. Extrapolations from controlled studies of rats, geese, mice, and apes under conditions high population density and over- and understimulation appear to be no longer accepted uncritically by responsible social and behavioral scientists. In particular, the literature of "overcrowding" appears to show not only significant differences between different cultural groups, but also that nonurban condition may not be as social and physical problem-free as they had been assumed to be.[18]

Nevertheless, our immediate concern is more with image and intuition than with reality. Despite the fact that densities and person-per-room have steadily declined in recent decades, levels of expectation and preference have risen as well. It would appear that as rural and agrarian experiences are part of the experiential baggage of lesser and lesser portions of the urban population the potential for romantic imageries of such ways of life are somewhat commensurately enhanced. The popularization of the more pointed and stridently critical ethological literature, coupled with literature of environmentalism that has been critical of large-scale urbanization, appears to reify many of these images.

As implied in the earlier discussion the fact of whether urbanism is intrinsically inimical to the best individual and collective spiritual and social interests of man may have as much to do with images as with the objective conditions of the city. A related matter may be that urban man may be more willing to abide by competition with nature than competition with his fellow man in which the rules of the game seem anything but constant or fair. Indeed, some of the rhetoric of the more strident antiurban environmentalism poses that the nonurbanized state is one in which man exists in a balanced and harmonious relationship to nature.

The Specialization-Community Paradox

A fourth root cause view of antiurbanism may be constructed from the specialization of labor-demise of community paradox, which posits that the increasing specialization of labor and the sense of community among urbanites are inversely related.[19] This perspective, which takes some of its inspiration from Durkheim, Wirth, and Redfield, suggests that need for a sense of community and commonality of purpose in highly interdependent urban society is made more difficult by the very process that creates that need. The paradox is that although specialization of labor heightens the degree the interdependency among city dwellers this specialization also creates "a bondage" of dependency. Urban man is increasingly dependent upon the specialties of his fellow urbanites for the bewildering array of skills and expertise required keeping his urban world functioning.

The social and psychological hidden price for the efficiency and economy of the specialization of labor is evident in the expanding dependency relationships of urbanites upon their fellow urbanites. One unobtrusive measure of this interdependency is evident in the list of telephone numbers (for doctors, lawyers, accountants, and repair men in various appliances and automobiles, etc.) upon which urbanites depend to maintain and facilitate these extensive relationships day by day. Furthermore, the urbanite is likely to find these relationships conducted increasingly at an impersonal and contractual level.

While "do-it-yourself-ism" may have its economic motivations it may also be seen as a means of lessening this dependency relationship and as a quest for self-sufficiency and relief from the routinization that often attend the specialization of labor. The effects of the increasing refinement of the division of labor are also evident in the "problem of community." In part this is reflected in the often discussed "interest communities" (in contrast to place-based communities),[20] wherein urbanites tend to form stronger attachments to their vocational communities (professional societies, labor unions, social and recreational organizations, etc.), than to the place-communities in which they reside. Moreover, place-based communities and neighborhoods may come to be composed more of individuals and families whose primary common bond is an income level sufficient to purchase a home in such areas.

Finally, the community-eroding effects of specialization are also evident within interest communities as well. The hyphenization of professional activities—physician-urologists, divorce-attorneys, ecologist-planners, corporate-accountants, and so on—may also represent subinterests and divergent professional values and norms that are corrosive of commonality of purpose in these types of communities as well.

The more typical expressions of the specialization community paradox are evident in the extensive social commentary, which finds concern and disillusionment with the impersonality, bureaucratized, and competitiveness of modern urban life and culture. While these expressions are generally indictments of modern industrial culture, that culture is largely embedded in the city.

The Social Control Paradox

Somewhat related to the above hypothesis is the paradox of social control. In coming together in cities to enjoy their opportunities and benefits urban man has had to increasingly subordinate autonomous choices to the concept of the commonwealth. While on its face this may appear as a rather axiomatic observation, it has undoubtedly provided additional fodder for sentiments of urban discontent. In part, this hypothesis is strongly related to the incongruity of attitudes such as "rugged individualism" and the "frontier ethic" in the American folk tradition with the complex legal, social, and administrative demands of urban life in America, compounded with the fact that American urbanization has been an experience of the rapid siphoning off of population from the rural and agrarian sectors of the society has resulted in the burden of civic socialization. Encompassed in the tradition of rugged individualism is the notion of freedom of action and individual expression and personal autonomy. Urbanism, in some contrast, places requirements upon individuals that demand subjugation to a complex set of laws that stipulate that individual behavior be compromised against the notion of the common good or public interest.

This hypothesis may also be expressed in an additional dualism: the contrast between what might be called "locational freedom" and "freedom of opportunity." The former notion is more expressive of nonurban contexts, particularly nomadic and agrarian. These contexts, while they present quite limited opportunities for occupational opportunity, cultural diversity and technological superiority, offer in their spatial remoteness, lower population levels and institutional simplicity a context that requires less social control over the social externalities of behavior. It would appear that this is in part the appeal of such disurban expressions as the agrarian commune and other "counterculture" social organization schemes. Even more succinctly, it is the operative mechanism in the so-called do-your-own-thing ethic. In contrast, the city offers great freedom of opportunity in terms of occupation and cultural diversity; but such opportunity is purchased at the cost of subordination to the rules and regulations of highly aggregated and diverse population complexes with concomitant extensive social externalities flowing from individual behaviors.

This view, of course, has its expression in other dimensions, at least one being the traditional aversion of Americans from socialistic and communistic social forms, both of which posit high degrees of individual subjugation to the stability and ends of the state. In sum the urban way of life paradoxically exacts a price for its greater opportunity structure and requires of its members a heightened awareness and subsequent willingness to abide the laws that attend of the notion of the greater good. But these awareness overlays upon the other anxieties of urban man the awareness that the welfare of the city (or at least those who are in a position to control it) takes precedence over individual welfare. Second, it places an additional burden upon urban man to find means to participate in the city's decision-making structure, to become more political.

The price of his liberty becomes not only "eternal vigilance" but also effective participation. In coming to cities man mingles his destiny and welfare with those of his fellow urbanities; the rugged individual, whatever his motives, is posed against the system.[21]

The Imperial City

The historical position of cities as engines of economic, political, and military power does not require any repetition here. Historically imperial power has always resided in cities and the ability for conquest and subjugation of any one power by another has necessitated the coup de grâce of urban annihilation. The power of cities and urbanized nations is no less demonstrable today than in earlier times, and part of the misgivings that urban man may have about cities may well be an anxiety that derives from the enormity of urban power.

From an intranational point of view this anxiety may derive from the realization that, whatever choices an individual may make as to lifestyle and location, the city with its enormous economic and political dominion in urbanized societies leaves the nonurbanite subject to the controls and whims of the urban sector. This power of the city has been evident in various forms: the conglomerate power of agribusiness over the small yeoman farmer, the domination of big over small business, the tenuous dependency of small single-industry communities upon the external control of corporate decision making, and the swallowing up of small independent communities by rampant urbanization. This theme has been put other ways, for example, that cities develop to a point beyond their responsiveness to controllability by their citizens. Such concerns are frequently cited as fundamental causes for the governmental fragmentation of urban regions and attest to the longevity of the Jeffersonian biases against urban massification.

Conclusion

The speculative nature of this discussion leaves a good deal open to question. Terms like urbanization and city obviously have different meanings and experiential referent to different people. While these subjective factors raise certain difficulties with attempts to test the hypotheses offered here with vigor and accuracy, the points raised here do indicate that other forms of survey instruments beyond those that have simply solicited living location preferences may well be useful in expanding this subject. The initial paradox offered above—that these sentiments persist in an atmosphere of continued urbanization—suggests that perhaps analysis of the attitudes of nonurban residents and new migrants to cities warrants deeper investigation. At least one international study, for example, has indicated that in terms of positive assessments of their person situations, "except for the most developed nations and Nigeria, urban residents rated themselves higher than did rural residents."[22] The city does not appear to have diminished significantly as a beacon of hope and possibility for the countryman, but it may be useful to know what causes disillusionment once he arrives.

In an earlier paper this author conducted a selected survey of children's literature and suggested that while the antiurban and rural-romantic themes prevalent in them may not be causes of later antiurban sentiment, they may serve to prepare children to accept uncritically antiurban themes in religious and other literatures.[23]

Anyone familiar with the literature of urbanization is aware that almost from the beginning cities have been both loved and loathed. Indeed it may be a characteristic of urban man that his quest for the good life constantly forces his expectations to raise, seeking the charges that are necessary to perfect his urban creation. The difficulties of doing so may well cause him to be wistful of more simplistic social orders.

The thesis offered above was that, through urbanization, man has increasingly acquired knowledge about himself and the universe, which has enabled him to take more rational charge of his future; but his same process has been attended by secularization and gelatinization. It was suggested that the process of urbanization has perhaps affected a disjunction between urban man and conscious and subconscious needs for guidance, meaning, continuity, and individuality. Perhaps an old Italian proverb sums it up: "He who forsakes the old way for the new knows what he has lost, but not what he will find."

Notes

Originally published as "The Roots of Anti-Urbanism," The Institute of Public and Urban Affairs, San Diego State University, January 1978.

1. E. Gordon Ericksen, "Some Historical Perspectives on the Anti-Urban Bias," in Jeffrey K. Hadden, Louis H. Masotti, and Calvin J. Larson (eds.), *Metropolis in Crisis* (Itasca, IL: Peacock, 1967), 176–177.
2. Ibid., 179–181.
3. Carl E. Schorske, "The Idea of the City in European Thought," in Oscar Handlin and John Burchard (eds.), *The Historian and the City* (Cambridge, MA: MIT Press, 1963), 99–114.
4. Leo Marx, *The Machine in the Garden: Technology and the Pastoral Ideal in America* (New York: Oxford University Press, 1964); Raymond Williams, *The Country and the City* (New York: Oxford University Press, 1973); Morton White and Lucia White, *The Intellectual Versus the City* (New York: Mentor, 1963); Iving Howe, "The City in Literature," *Commentary* (May 1971): 61–68.
5. Morton White, "The Philosopher and the Metropolis in America," in Werner Z. Hirsch (ed.), *Urban Life and Form* (New York: Holt, Rinehart and Winston, 1963, 1963), 81–96.
6. Ibid., 88–89.
7. Paul Wheatley, *The City as Symbol* (London: H.K. Lewis, 1967).
8. *Time Magazine* (December 25, 1972): 13. The study referred to is entitled *The State of the Nation*, William Watts and Lloyd A. Free (eds.), published by the Potomac Associates, 1972, 81–83.
9. As reported in Irving Kristol, "An Urban Civilization," *Horizon* (Autumn 1972): 36.

10. Beggs and Copeland, "Campus Opinion," *San Diego Evening Tribune* n.d.
11. Florence Ladd, "Black Youths View Their Environments: Some Views on Housing," *Journal of the American Institute of Planners* 38 (2) (March 1972): 111.
12. Gunilla Lonoeus, "80% of Youth Say 'No' to City," *Sweden Now* (July/August 1971): 24–25.
13. Joyce O. Hertzler, *The Social Thought of Ancient Civilizations* (New York: McGraw-Hill, 1936), 298–306. See also J. Ellul, *The Meaning of the City* (Grand Rapids: Eerdmans, 1970); Harvey Cox, *The Secular City* (New York: Macmillan, 1965); Leo Marx, "Pastoral Ideal and City Troubles," in *The Fitness of Man's Environment* (Washington, DC: Smithsonian Institute Press, 1968). The New Testament, while not as pointedly antiurban as the Old Testament, offers little recommendation of city life.
14. Jeffrey K. Hadden and Josef J. Barton, "An Image That Will Not Die: Thoughts on the History of Anti-Urban Ideology," in Louis Massotti and Jeffrey Hadden (eds.), *The Urbanization of the Suburbs* (Beverly Hills: Sage, 1973), 87–88.
15. Cox, *The Secular City*, and Michener, *The Source* (New York: Random House, 1965).
16. Cox, *The Secular City,* p. 30.
17. Robert Ardrey, *The Territorial Imperative* (New York: Atheneum, 1966); Konrad Lorenz, *On Aggression* (New York: Harcourt, Brace & World, 1966); Desmond Morris, *The Naked Ape* (New York: McGraw-Hill, 1967). See also Edward T. Hall, *The Hidden Dimension* (Garden City, NY: Doubleday, 1966).
18. Cf., for example, John Cassel, "Health Consequences of Population Density and Crowding," in Robert Gutman (ed.), *People and Buildings* (New York: Basic Books, 1972), 249–268.
19. Lawrence Haworth, *The Good City* (Bloomington: Indiana University Press, 1963), 19.
20. Melvin Webber, "Order in Diversity: Community Without Propinquity," in Lowden Wingo (ed.), *Cities and Space* (Baltimore: Johns Hopkins Press, 1963), 23–54.
21. C.W. Griffin, "The Frontier Heritage of Urban America," in Melvin I. Urofsky (ed.), *Perspectives on Urban America* (Garden City, NY: Doubleday, 1973), 23–42.
22. Claude S. Fischer, "Urban Malaise," *Social Forces* 52 (2) (December 1973): 224.
23. James A. Clapp, "City Mouse/Country Mouse: Planning and the Wellsprings of Anti-Urbanism," Delivered at the National Conference of the American Institute of Planners, Atlanta, Georgia, October 23, 1973.

PART 2

Antiurbanism in History and Literature

CHAPTER FOUR

*Pastoral Ideals and City Troubles**

LEO MARX

> She told him about her childhood on a farm and of her love for animals, about country sounds and country smells and of how fresh and clean everything in the country is. She said that he ought to live there and that if he did, he would find that all his troubles were city troubles.

The woman whose opinions are being reported here is Betty, the robust, beautiful heroine of Nathanael West's macabre fable of modern American life, *Miss Lonelyhearts*. She is offering them to the protagonist, the writer of an advice-to-the-lovelorn column, Miss Lonelyhearts himself, who is neurotically obsessed with the anguish of his correspondents. And he momentarily assents, as many of us would, to Betty's plausible argument. She exaggerates, to be sure, yet who would deny that a great many of our troubles are city troubles? What does give us pause, however, is the notion that we can cope with them by retreating to the country. How shall we take this familiar idea? We know that it is deeply implanted in American culture, and especially in our literary culture. We know that American writers, from the beginning of a distinct national literature, have been fascinated by the theme of withdrawal from a complex, relatively "advanced" civilization to a simpler, more natural environment. This movement in space typically has served to represent a movement of mind and spirit—a quest for a new and happier way of life. And even in the twentieth century, when the theme might be thought to have lost its relevance, it has in fact retained its hold upon the imagination of many of our leading writers. Why? What does it signify? What bearing can it possibly have upon the problems of our urbanized society? My aim here is to answer these questions, and to suggest some ways in which the answers may be useful to those who plan the development of our physical environment.

But as a student of American culture, and one who has been concerned with the interplay between literary and extra-literary experience, I recognize that most attempts to trace the mundane consequences or implications of imaginative writing have been unsatisfactory. The crux of the difficulty is the need to make connections between two kinds of discourse. In poetry and fiction the controlling context is imagistic and metaphoric, and when we attempt to translate its meaning into everyday, practical language we all too often flatten the intricate, multidimensional structure of image, thought, and feeling; by reducing literary language to merely logical, discursive statements, we lose touch with precisely that affective power that is, after all, the distinctive property of literature—its reason for being. To name this difficulty, however, is to suggest why the present enterprise could be worthwhile. Because imaginative literature remains one of our most delicate and accurate means of joining ideas with emotions, public with private experience, I believe that it can provide insights into the relations between mind and environment that are unavailable elsewhere. I want to show that the literary landscape, properly understood, could help us in planning the future of the actual landscape. I do not propose, of course, that literary works can be made to yield a blueprint or, for that matter, any specific, tangible features of a physical plan. But I do believe that they can help us sort out, clarify, and reorder the principles that guide (or should guide) the planners.

No one needs to be reminded that imaginative writing, especially in the modern era, is a storehouse of ideas and emotions that men have attached to the landscape. In the American consciousness, as D.H. Lawrence observed long ago, the spirit of place is particularly strong. I want to begin, therefore, with an ideal type of a familiar symbolic landscape—one that recurs everywhere in our native literature. This terrain characteristically has three sectors: a community (village, town, or city); a partly developed middle ground, neither urban nor wild; and a wilderness. But this imaginary countryside does not serve our writers merely as a backdrop or setting. In the best known American fables—I am thinking, for example, of Thoreau's *Walden*, Melville's *Moby Dick,* and Mark Twain's *Huckleberry Finn*—the symbolic landscape is inseparable from the action or narrative structure, which may be divided into three movements: the retreat, the exploration of nature, and the return.

First, then, the retreat. The action begins with the hero-narrator's withdrawal from a relatively complex, organized community from which he is alienated. Here life seems to be dominated by an oppressively mechanistic system of value, a preoccupation with the routine means of existence and an obliviousness of its meaning or purpose. Here, Thoreau says, men have become the tools of their tools. Unable to relate his inward experience to his environment, the narrator retreats in the direction of nature.

In the second, or central, movement he explores the possibilities of a simpler, more harmonious, way of life. At some point, invariably, there is an idyllic interlude when the beauty of the visible world inspires him with a sense of relatedness to the invisible order of the universe. During this episode, which can only be described as a moment of religious exaltation, he enjoys an unusual feeling of peace and harmony, free of anxiety, guilt, and conflict. But the possibilities of a life beyond the borders of ordinary society prove to be limited, and two characteristic kinds of episode help to define those limits.

In one, which may be called the interrupted idyll, the peace and harmony of the retreat into the middle landscape is shattered by the sudden, often violent intrusion of a machine, or of a force or person closely associated, in the figurative design, with the new industrial power. (Recall the scene in which the shriek of the locomotive destroys Thoreau's revery at Walden Pond; or the episode when Ahab's violent declaration of purpose, which he associates with mechanized power, follows Ishmael's pantheistic masthead dream; or the decisive moment when the steamboat smashes into the raft in *Huckleberry Finn*.) The second characteristic limiting episode occurs when the narrator's retreat carries him close to or into untouched, untrammeled nature, and though his exposure to the wilderness often proves to be a spiritual tonic, evoking an exhilarating sense of psychic freedom, it also arouses his fear. For he soon comes to recognize that an unchecked recoil from civilization may destroy him—either in the sense of extinguishing his uniquely human traits or in the quite literal sense of killing him. He discovers, in short, that there are two hostile forces that impinge, from opposite sides of the symbolic landscape, upon the gardenlike scene of his retreat: one is the expanding power of civilization, and the other is the menacing anarchy of wild nature.

These insights lead, however indirectly, to the third and final phase of the action: the return. Having discovered the limited possibilities of withdrawal, above all its transience, the narrator now returns, or seems to be on the point of returning, to society. But the significance of this movement, which is also the ending of the work, is clouded by ambiguity. Has the hero been redeemed? Is he prepared to take up, once again, the common life? What is he able to bring back, as it were, from his exploration of the natural environment? Though he apparently acknowledges that society is inescapable, he usually remains a forlorn and lonely figure. Our most admired American fables seldom, if ever, depict a satisfying, wholehearted return, and in the closing sentences of one of them—*Huckleberry Finn*—the protagonist already has begun a new retreat, as if to suggest an unending cycle of withdrawal and return.

So much, then, for the design of the symbolic landscape. I propose to show that it is an embodiment of a more or less coherent view of life, a conception of the relations between imagination and reality, which may be called a peculiarly American version of romantic pastoralism. Before attempting to describe the viewpoint and its contemporary implications, let me briefly consider its specifically pastoral, distinctively American, and romantic components.

The "psychic root" of this thematic design, perhaps of all literary pastoralism, is the impulse to retreat from a complex society in search of happiness and virtue. In Western literature the theme can be traced to the work of Theocritus and Virgil, but in fact we all know it at firsthand. It is the familiar urge, in the face of civilization's growing complexity and power, to "get away"—to leave a complex world (traditionally associated with the royal court and city) and begin a new life in a simpler environment (traditionally associated with the actual rural landscape). The pastoral element of the design, then, lends expression to this centrifugal impulse; it turns upon the contrast between two styles of life, one sophisticated and the other simple, one identified with a relatively "advanced" society, the other with a life "closer to nature." The continuing appeal of pastoralism evidently derives from the universality of the conflict represented by the two physical environments, and if there is a single device that may be considered a constant feature of the mode, it is the symbolic landscape that has been used to figure forth that conflict from Virgil's time to that of Robert Frost, Ernest Hemingway, and William Faulkner.

To appreciate the special affinity between the pastoral mode and the American consciousness, we have only to recall the symbolic topography invented by Virgil. We all remember Arcadia, the ideal site of harmony, beauty, and material sufficiency that chiefly engages Virgil's attention.

But we tend to forget the extent to which this earthly paradise derived its charm from the two contrasting kinds of terrain upon its borders. In the first eclogue Virgil insists upon the encroaching presence both of Rome—locus of imperial power, authority, and repression—and of the bare rocks and marshland that epitomize unimproved, inhospitable, infertile, wild nature. Pastoralism may be regarded as an ecological literary mode, its purpose being to mediate between the claims of these two conflicting yet inescapable human environments: one associated with man's biological origins, the other a product of technological change and sociocultural evolution. When the pastoral ideal is pictured as a middle landscape located between the extremes of wildness and overcivilization, it is easy to see why it lent itself, beginning in the Age of Discovery, to interpretations of life in the New World. Here, in place of an imaginary Arcadia, was the utopian promise of the new colonies, with the old world to the east, realm of sophistication, power, and history, and the whole reach of the North American wilderness to the west. It is not surprising, under the circumstances, that the transit of Europeans to America often was conceived, like the good shepherd's retreat to an ideal, green pasture, as a movement toward a new, simpler, and happier way of life.

The crucial distinction, then, between American and traditional versions of pastoral, is the new realism that was imparted to the ideal by the new world situation. Before the Renaissance, poets had habitually depicted a dreamland. But in Shakespeare's time the symbolic landscape that had for so long been considered a poetic figure suddenly acquired a real geographic location. Now the pastoral ideal was taken seriously, with a novel literalness, as a social and political possibility, and its temporal location was shifted from the golden past to the utopian future. In America, by Jefferson's time, it had acquired political as well as geographical reality. When the authors of the Declaration of Independence rephrased John Locke's enumeration of the rights for whose protection governments are instituted, replacing his "life, liberty and property" with "life, liberty and the pursuit of happiness," they in effect transferred the ancient pastoral dream of human possibilities from its traditional literary context to an actual political context. No wonder the enemies of the third President of the United States called him a poet and dreamer! In formulating the goals of the Republic, Jefferson subordinated material well-being, national wealth, and power to what nowadays would be called the overall "quality of life."

That the American public responded favorably to the pastoral idiom of the Jeffersonians seems beyond dispute. During the nineteenth century the image of a green garden, a rural society of peace and contentment, became a dominant emblem of national aspirations. In the general culture the image of the garden served to blend the ideals derived from literary pastoralism and from Christianity. Only the most astute grasped the contradiction between the kind of society that Americans said they wanted and the kind they actually were creating. While the stock rhetoric affirmed a desire for a serene, contemplative life of pastoral felicity, the nation's industrial achievements were demonstrating to all the world its tacit commitment to the most rapid possible rate of technological progress, and to an unlimited build-up of wealth and power. This is the conflict of value dramatized by the interrupted idyll, the episode in which a machine suddenly destroys the tranquillity of an asylum in nature.

Before this time, however, the attitudes born of international romanticism also had been assimilated to the native version of the pastoral design. To elucidate the complicated and obscure relations between the romantic vision and the pastoral mode is beyond the scope of my subject. Suffice it to say, here, that under the influence of the romantics the pastoral retreat into nature took on a far more explicitly metaphysical, quasi-religious significance. By Wordsworth's time the natural landscape had become a repository for those ultimate values formerly attributed to the Christian deity. As Emerson put it, nature (which he tended to represent by landscape images), had become for his generation "the present expositor of the divine mind." At the same time in the "high culture," the machine was becoming a dominant symbol for the impersonal, squalid, and inhumane world of the new industrialism, so that the movement in the direction of nature now could be depicted as a melodramatic withdrawal from a cold, mechanized city into a warm, living, spiritually nurturing countryside. In the romantic era, native pastoralism acquired new vitalistic sanctions. If the retreat to the countryside made possible a simpler, more harmonious earthly existence, it was because it provided closer access to divine sources of order, meaning, and purpose.

So much, then, for the classic, American version of the pastoral design. That it engaged the attention of our writers in a period when a vast population was moving into a seemingly prehistoric landscape hardly is surprising. But it is more difficult to account for its continuing hold upon the literary imagination in the twentieth century. Again and again, in the work of writers like Frost, Fitzgerald, Hemingway, and Faulkner (to name only a few famous examples), we find a similar preoccupation with the pastoral impulse, that is, with a retreat from urban society toward nature as the outward expression of a quest for happiness, order, and meaning. Yet anyone familiar with the work of these writers would agree, I believe, that they are not sentimentalists; we cannot imagine them seriously entertaining the illusion, cherished by Nathanael West's Betty, that we can solve our city troubles by moving to the country. Their work does not, in other words, encourage us to believe that the recovery of a rural style of life is a genuine alternative to life in our intricately organized, urban, industrial society. But for what purpose, then, do they continue to employ the pastoral design? Why does it engage the attention of so many of our best writers and, presumably, the audience who admire their work? What, in short, is the significance of the design?

To answer the question, I will consider examples from the work of Robert Frost and Ernest Hemingway. Of all modern American writers, Robert Frost belongs most directly in the line from Virgil and the romantic pastoralism of Wordsworth and Emerson. It is significant that he placed, as the first poem in the *Complete Poems*, a brief and deceptively slight invitational lyric, "The Pasture." There he invites us to leave the house of everyday life and move out toward nature.

> I'm going out to clean the pasture spring;
> I'll only stop to rake the leaves away
> (And wait to watch the water clear, I may):
> I sha'n't be gone long.—You come too.
>
> I'm going out to fetch the little calf
> That's standing by the mother. It's so young
> It totters when she licks it with her tongue.
> I sha'n't be gone long.—You come too.

Like most of Frost's work, the poem may be taken in two ways, either in the plainest sense, for the pleasure of reference, or for its extended meaning. In this case we are also being invited into a poetic world, an ideal pasture where the writer will clear a channel to a hidden source of renewal and creativity. Raking away the clutter of dead leaves and nurturing the just born calf are actions that suggest how much—and how little—he expects of the retreat.

The landscape sketched here is the symbolic landscape of Frost's memorable lyrics. As in earlier versions of the pastoral design, this topography is divided into three sectors: a community, a middle terrain or pasture, and beyond that the dark woods and desert places. When

Frost occasionally looks directly at organized power, he too is likely to represent it by technological imagery (as in "A Brook in the City" or "The Egg and the Machine"), but the typical Frost lyric turns upon a moment after the speaker already has turned away from the urban-industrial environment. The poet's subject is retreat, and in the opening line of "Directive" he encapsulates the root impulse of native pastoralism: "Back out of all this now too much for us." How many modern American novels and poems begin with variants of this impulse! The theme is retreat—both what it promises and what it threatens—and it carries the speaker into a middle ground—for Frost it is likely to be a meadow with a brook at its center—where the water wells up from a savage source, offering the hope that we might "Drink and be whole again beyond confusion." But withdrawal into nature has specific limits in space and time, and Frost is careful to insist, always, that it must end with a return to the common life. "I sha'n't be gone long."

By now it should be evident that the pastoral motif as used by Frost is largely drained of the literal meaning it had acquired in Jeffersonian political discourse. Here the movement outward from society toward nature has little to do with the practical superiority of rural ways, and in a poem like "New Hampshire" Frost ends by mocking that idea. The concluding lines, where the speaker faces a choice between being "a prude afraid of nature" or a "puke," that is, between a prudish New England rustic or a New York (Freudian) smartalec, he says:

> Well, if I have to choose one or the other,
> I choose to be a plain New Hampshire farmer
> With an income in cash of say a thousand
> (From say a publisher in New York City).
> It's restful to arrive at a decision,
> And restful just to think about New Hampshire.
> At present I am living in Vermont. [~]

And in a letter to William S. Braithwaite, Frost made clear his skepticism about sentimental pastoralism.

> I kept a farm, so to speak, for nearly ten years but less as a farmer than as a fugitive from the world that seemed to me to "disallow me." It was all instinctive, but I can see now that I went away to save myself and fix myself before I measured my strength against all creation. I was never really out of the world for good and all.[1]

The significance of Frost's retreat, in short, is not primarily social or political, but rather psychological or metaphysical. Its value is inward. What impels the speaker is a yearning for an indefinable value, order, meaning—a sense of relatedness to that Wordsworthian "something" that is unavailable in the social environment. Following the romantics, Frost is tempted by the notion that natural facts, properly perceived, can be made to yield a surrogate for the moral or metaphysical coherence formerly expected from Christian revelation. And so, again and again, in poems like "The Most of It" or "Mowing," he seizes upon a particular natural fact, suggesting the tantalizing possibility that he may be able to wrest from it a moral or transcendent meaning. In "Mowing" the speaker is working in a soundless pasture beside the wood, when it occurs to him that his relation with nature, figured by the whispering sound of the scythe, can be captured in a statement. "What was it it whispered?" he asks, but in the end he characteristically retreats from that invasion of the realm beyond the visible, where Nature is an embodiment of ultimate value, to a mediating middle ground: "The fact is the sweetest dream that labor knows."

Not only is the "content" of Frost's poetry controlled by the pastoral design, but his explanation of the creative process, and of the function of poetry, conforms to the same pattern. The inception of a poem is an impulse similar to the retreat, or what Freud might have called an enactment of the pleasure principle. "It begins in delight," Frost says, "and ends in wisdom. The figure is the same as for love." As he describes it, there is a similar reaching out for gratification, a similar arrest of the centrifugal motion, checked in this case by the requirements of form, and then a denouement comparable to the hero's return:

> It begins in delight, it inclines to the impulse, it assumes direction with the first line laid down, it runs a course of lucky events, and ends in a clarification of life—not necessarily a great clarification, such as sects and cults are founded on, but in a momentary stay against confusion.[2]

With Frost, then, the pastoral design is more than a convenient device for structuring a work of art. It figures the rhythm of consciousness itself; it is a landscape of mind. Moreover, Frost's popularity—and he is beyond question the modern American poet with authentic gifts who has the largest audience—would seem to argue the universal appeal of the design. His most popular poem, which has been subjected to endless critical explication, reprinted in mass circulation magazines, and repeatedly anthologized, is "Stopping by Woods on a Snowy Evening." It would be difficult to imagine a more complete statement of the theme, and familiar as the lines are, it is revealing to reconsider them with the pastoral design in mind.

> Whose woods these are I think I know
> His house is in the village though;
> He will not see me stopping here
> To watch his woods fill up with snow.

The complex institutional world has been left behind, and though we are made to feel how it impinges on the countryside, in the property owner's invisible presence, the speaker's attention is drawn to the landscape. The falling snow obliterates details, harmonizes the scene, and provides a receptive field for his meditation. (As in "Directive": "Back in a time made simple by loss of detail.") Yet his withdrawal is far from complete.

> My little horse must think it queer
> To stop without a farmhouse near
> Between the woods and frozen lake
> The darkest evening of the year.
>
> He gives his harness bells a shake
> To ask if there is some mistake.
> The only other sound's the sweep
> Of easy wind and downy flake.

Now the sense of being in the precarious middle, having to mediate the claims of two environments, becomes acute. The horse, trained to a workaday routine, would pull the speaker back to the daily round. But the speaker is transfixed by the serenity and beauty in the sphere of nonhuman nature, and the enticing, barely audible whisper of the snow, like the sound of the scythe in "Mowing," suggests some obscure fulfillment or incipient transcendence. The temptation to keep going is strong. The poem ends:

> The woods are lovely, dark and deep,
> But I have promises to keep,
> And miles to go before I sleep,
> And miles to go before I sleep.

In spite of the loveliness of the woods, with all that they imply of soothing release, the speaker turns away, as if aware that to continue his retreat is to court the ultimate simplification—a total merging into dark otherness, a deathlike loss of self. The repetition of the final line underscores the precariousness of this resolution, but in the end he has turned back to the obligations of common life.

Turning now to the work of Ernest Hemingway, we often find a strikingly similar moral landscape. Organized society is identified with organized violence, often with a brutal war—mechanical, meaningless killing. Again, the action originates in something like the pastoral impulse to get away; the hero has been wounded, physically or psychically or both, and he opts out—declares a separate peace—retreating from the impersonal cruelty toward a simpler life in a natural setting. The retreat may take the form of an African safari, a fishing trip in Michigan or Spain, but its true object, as in Emerson, Thoreau, or Frost, is psychic and moral renewal. But consider a specific example.

"Big Two-Hearted River," the final story in Hemingway's first book, *In Our Time*, has had a strange history. At first many readers, including some who were intrigued by it, thought the story was pointless. Once F.Scott Fitzgerald and Dean Gauss taunted Hemingway for "having written a story in which nothing happens."

"Big Two-Hearted River" is an account of a two-day fishing trip that Nick Adams takes in Michigan. In the course of the story Nick leaves the train that has brought him to the country, hikes overland to a meadow where he pitches his tent, eats supper, and goes to sleep; the next day he goes fishing, catches some fish, and decides not to fish in a swamp. That's all. He meets no one, and there is no significant action. The whole tale is told in simple, declarative sentences, constructed with a fastidious attention to detail and in a seemingly calculated monotonous rhythm. Here is an example:

> There was no underbrush in the island of pine trees. The trunks of the trees went straight up or slanted toward each other. The trunks were straight and brown without branches. The branches were high above. Some interlocked to make a solid shadow on the brown forest floor. Around the grove of trees was a bare space. It was brown and soft underfoot as Nick walked on it.[3]

Since its publication in 1925, a number of critics—notably Edmund Wilson, Malcolm Cowley and Philip Young—have uncovered certain of the story's unstated themes. They have noticed that it belongs to a chronological sequence of Nick Adams stories in the book, which has a degree of thematic unity. Nick has suffered a traumatic wound in the war, and now he has come back to Michigan to recuperate. On careful inspection, it becomes evident that the central action lies beneath the surface, in Nick's mind, and that the numbed, almost lobotomized prose is an index of his effort to repress his panicky emotions.

But the implications of "Big Two-Hearted River" become even more evident when we examine its symbolic landscape. Here is the way the story begins:

> The train went on up the track out of sight, around one of the hills of burnt timber. Nick sat down on the bundle of canvas and bedding the baggage man had pitched out of the door of the baggage car. There was no town, nothing but the rails and the burned-over country. The thirteen saloons that had lined the one street of Seney had not left a trace. The foundations of the Mansion House hotel stuck up above the ground. The stone was chipped and split by the fire. It was all that was left of the town of Seney. Even the surface had been burned off the ground.[4]

The great world from which Nick has withdrawn is represented only by the train, moving out of sight, and by the unexplained fire that has obliterated this outpost of civilization. The surrounding country is a burned-out wasteland. For a time, Nick walks over blackened earth with no green vegetation; even the grasshoppers have turned black. The idea that preoccupies him is getting away. "He felt he had left everything behind, the need for thinking, the need to write, other needs. It was all back of him." Or, in Frost's words, "Back out of all this now too much for us." Later he reaches the place where the fire ended, and the country turns green. He walks along the river in the hot sun. When he finally selects a spot to make camp, it is a meadow on the shore of the river, and on the other side there is a dark swamp. The tram and the burned out town and war-decimated Europe are behind him. Like Thoreau at Walden, his aim is to reduce life to its simplest elements. He cooks his dinner, makes his bed, and gets ready for sleep.

> Nick was happy as he crawled inside the tent. He had not been unhappy all day. This was different though. Now things were done. It had been a hard trip. He was very tired. That was done. He had made his camp. He was settled. Nothing could touch him.

> It was a good place to camp. He was there, in the good place. He was in his home where he had made it.[5]

Nick's camp is a psychological middle landscape, and the next day, when he goes fishing, his mind starts to work. But he cannot bear too much emotion. When he loses a big fish he becomes overexcited, feels sick, and decides not to "rush his sensations." Throughout the detailed account of the fishing, Hemingway reminds us several times of the swamp across the river. Its presence makes Nick uneasy. Toward the end of the story it becomes his preoccupation, like the dark woods that transfix the speaker of Frost's poem.

> He did not feel like going on into the swamp. He looked down the river. A big cedar slanted all the way across the stream. Beyond that the river went into the swamp.
>
> Nick did not want to go in there now. He felt a reaction against deep wading with the water deepening up under his armpits, to hook big trout in places impossible to land them. In the swamp the banks were bare, the big cedars came together overhead, the sun did not come through, except in patches; in the fast deep water, in the half light, the fishing would be tragic. In the swamp fishing was a tragic adventure. Nick did want it. He did not want to go down the stream any further today.[6]

And then, after Nick cleans his fish, and washes them, the story ends abruptly and, as many readers have testified, enigmatically.

> Nick stood up on the log, holding his rod, the landing net hanging heavy, then stepped into the water and splashed ashore. He climbed the bank and cut up into the woods, toward the high ground. He was going back to camp. He looked back. The river just showed through the trees. There were plenty of days coming when he would fish the swamp.[7]

And miles to go before I sleep! What is striking here, apart from the similarity to Frost's version of the design, is the close correlation between the external landscape and the pattern of Nick's inner life—the structure of his feelings. First, the burned-over land, identified with machines and war, and with Nick's anxious sense of threatening, repressive, wounding forces; then the camp in the meadow, a good safe place midway between the world of collective imperatives; and, finally, raw nature, represented by the third sector of the landscape, the swamp identified with darker, impulse-ridden, unknown life that is at once attractive and frightening. Nature in Hemingway's world is both benign and menacing—it is, like the river, two-hearted. Variations of this design recur elsewhere in his work, as they do in the work of other gifted American writers of our time. But here the story will suffice as a representative embodiment of a peculiarly American version of post-industrial romantic pastoralism.

With these examples from the work of Frost and Hemingway in view, it should be evident that the pastoral design does not embody an unqualified affirmation of the initial retreat toward nature. The movement in the direction of a simple, preindustrial setting does not arrive at an alternative, in any literal sense, to the complex world we inhabit. What it does offer, however, is a symbolic structure of thought and feeling, a landscape of mind in which the movement in physical space corresponds to a movement in consciousness. The literary topography is built, in other words, on a subjective model. The succession of contrasting spatial images (town, meadow, swamp) provides a vocabulary for expressing a sequence of feelings—feelings we ordinarily would regard as irreconcilable. A typical starting point is our ambivalent attitude toward the urban-industrial environment. We are simultaneously repelled and captivated by it; we may feel a strong impulse to escape from it, but we recognize that it is finally inescapable. By deploying these contradictory feelings in literary space the pastoral design enables us to sort them out, and to impose a degree of order upon them. To indicate how this happens, I shall review the tripartite structure of the design: the retreat, the exploration of the limits of nature, and the return.

The retreat from the complex world has both a negative and a positive aspect. The negative aspect is escapist. It expresses a revulsion against the more unpleasant features of the urban-industrial landscape: the ugliness, the noise, the poisoned air, the chaotic overabundance of stimuli, the symptoms of social disorganization, and the general impression of incoherence and individual powerlessness. Since Carlyle's time this environment often has been represented in literature by the image of a vast machine. As Lewis Mumford puts it, the word "machine" may be used to represent the dominant forces in the world today.

> Most of the creative forces in our time have been canalized into the Machine, a systematic organization of scientific discovery and technical invention that, under the pressure of excessive pecuniary

> gains and exorbitant political power, has transformed the entire existence of the Western World. The insensate dynamism of this mechanical organization with no goals but its own ceaseless expansion and inflation, has broken down the continuities of history.[8]

Whether or not it accurately represents the state of the "Western World" in our time, the image of the Machine as an emblem for a system with "no goals but its own ceaseless expansion and inflation" does express an attitude toward industrial society that permeates modem literature. It implies that the course of contemporary history is largely, perhaps irresistibly, determined by the course of technological development. And however much we may disapprove of this fatalistic idea, it does seem to be confirmed by the fact that when technical skill makes possible a flight to the moon, the building of an H-bomb or a supersonic jet, our society seems invariably to follow the lead of technological innovation. Given a world dominated by such a machine, in any case, the pastoral impulse to withdraw (or to "drop out," in the idiom of alienated youth), is an impulse to recapture a human situation as it might be imagined to exist beyond, or to have existed anterior to, our intricate technological order. The withdrawal of the pastoral hero in effect repudiates the assumption of Western culture that man is or can be wholly separated from nature, and that the environment exists chiefly as a source of raw material for the satisfaction of our unique needs. At the outset, accordingly, the hero does seem to deny any possibility of locating worthy purpose, meaning, or value within the collective existence of which the machine is our cardinal symbol.

The positive aspect of the retreat, on the other hand, may be described as a tribute to the pleasure principle. It expresses a desire to achieve felicity through a simplification of living that restores priority to basic instinctual gratifications. It is a search for precisely those qualities of life that our urban environment allegedly fails to satisfy. It would be useful, therefore, as a way of understanding the shortcomings of urbanism, to make a careful study of the satisfactions that writers of pastoral continue to identify with retreat to the natural landscape. I shall return to that proposal. But the point here is that the impulse to escape from our complex environment may prove to be regressive or progressive, depending upon what happens during the next stage: the exploration of nature.

The first thing to be said about the "return to nature" in our sophisticated pastoralism is that it avoids, or at least masks, the conventional romantic claim for the superiority of a rural or wilderness lifestyle. If the retreat can be salutary, it is not because it provides access to a mysterious, divine, or absolute principle inherent in the natural landscape. Our best writers are in remarkable agreement on this elusive metaphysical issue. Accordingly, they do not provide much comfort for those who would have us deal with city troubles by refurbishing preindustrial institutions. In other words, they recognize the irreversibility of history. But this is not to deny that they present the symbolic return to "nature" as a source of real satisfactions. Their work indicates that it can be just that. Again and again they show us that withdrawal from society in the direction of nature makes possible moments of emotional release and integration, a recovery of psychic equilibrium comparable to the release of repressed feelings in dreams or psychotherapy. This fact, to which the record of Western religion and literature abundantly testifies, imparts a degree of authenticity to the idea of a valuable "return to nature." At the same time, sophisticated writers of pastoral are virtually unanimous in their emphasis upon the limited value of such withdrawals from the world. Retreat is useful only if temporary. It does not—cannot—satisfy the hero's longing for a permanent alternative to our social environment. What it does provide, in Frost's telling phrase, is a "momentary stay against confusion." If unchecked, however, the pastoral impulse can lead to disaster. It leads the protagonist into the dark woods, or that swamp of instinct and uncontrolled feeling where, as Hemingway's hero fears, the fishing will be tragic.

What requires emphasis, then, is that while the first stage may seem to sanction the impulse to escape the machine of modern history, the second stage discloses the necessarily individualistic, transient character of the satisfactions that such an escape provides. Contrary to the connotations usually attached to the word, pastoralism reveals the inadequacy of the retreat to nature as a way of solving social and political problems. (The fate of William Faulkner's pastoral hero, Ike McCaslin, is perhaps our most eloquent testimony on this point.)

Thus the recurrent episode of the interrupted idyll has served to convey our writers' sense of disenchantment, however inchoate, with the promise of individual redemption that our literary culture took over from the radical antinomian strain in native protestantism. The retreat to nature represents moments of integration but, as Melville warned, "what plays the mischief with the truth is that some men will insist upon the universal application of a temporary feeling or opinion."

Hence the return. In the end, however equivocal the denouement may seem, the pastoral figure characteristically has turned back toward the world defined by the machine. Having discovered that the retreat can provide only a "momentary stay," he acknowledges that his true home is, after all, society. If the endings of our pastoral fables generally are unsatisfactory, if they seem to place the protagonist in equivocal, self-contradictory postures, it is largely because of the seemingly insoluble dilemma in which he has been put. How can he carry back into our complex social life the renewed sense of possibility and coherence that the pastoral interlude has given him? None of our writers has been able to find a satisfactory answer to this question.

At the outset I suggested that our imaginative literature, and particularly those works that embody the pastoral design, might be of some use—at least as a source of guiding principles—to those who plan the development of the physical environment. Yet there is a paradox here, for the significance of the design, as we have seen, is primarily subjective. It refers chiefly to the inner, not the external, landscape. The topographical imagery in our twentieth-century pastoral fables must be understood as a metaphoric representation of a landscape of consciousness. It would be a serious mistake, however, to conclude that the design therefore is irrelevant to the problems of the actual landscape. On the contrary, literary criticism insists that a powerful figurative relationship of this kind is never merely decorative or illustrative. It is not a one-way channel of meaning. If our writers consistently employ a pattern of landscape imagery, and if readers understand and lend assent to it—and, to repeat, the works that embody the pastoral design do seem to have a special appeal for a contemporary audience—then we must assume that the vehicle (in this case, the entire topographical design) is an indispensable feature of the total aesthetic result. So far as the design is convincing, the landscape imagery contributes to that conviction. Aesthetic success confers a kind of validity upon the pattern. What valid principles, then, can planners derive from the view of life inherent in this body of literature?

The most obvious inference is that more attention be paid to the subjective and in large measure traditional, aesthetic, or symbolic significance that our culture attaches to images of landscape—urban, rural, and wild. (On first looking into the literature of planning, the cultural historian cannot help being impressed by the lack of allusion to the centuries-old accretion of meaning that clings to our dominant topographical images.) The continuing hold of pastoralism upon the literary imagination in this urban-industrial age is but one measure of the power of such images. A specific measure, already mentioned, is that we undertake a thorough, precise, analytic inventory of the satisfactions that men have derived, or have claimed to derive, from various features of the landscape. Such a survey ultimately would require the collaboration, in addition to planners and literary scholars, of art historians, urban and rural sociologists, and psychologists. One aim of the inventory would be to sort out the kinds of satisfaction, real and illusory, that are associated with the pastoral retreat. Decisive here is the sense of repose, renewal, and sensual gratification identified with withdrawal from the city to a more natural environment. We know that some of these pleasurable feelings derive as much from what is missing as from what is actually present in the extra-urban setting. In the literary retreat, the pastoral figure's mind is released from the nagging responsibilities of a complex social life, the flood of conflicting stimuli, and the painful omnipresence of history itself. Some of his satisfaction, however, actually derives from the natural landscape, either from its specific physical attributes (the fresh air, the greenness, the color of flowers, etc.), or from its psychological and associative attributes. A natural setting, if only because it is less cluttered with man-made objects, provides a more hospitable field for the projection of his feelings. It thereby enhances, if only momentarily, his sense of his own power and importance. But a vital element here is the residuum of teleological modes of thought—the tendency to identify the seeming orderliness of the natural landscape with the hypothetical design and purpose of the cosmos. Whether these ideas are objectively "true" or not, they do in some measure control our responses to the physical environment, and planners might profitably make use of more information about them.

Another principle suggested by literary pastoralism is the importance of diversity in physical settings—the need to preserve the distinctness of the three spheres of our environment: the city, the rural countryside, and the wilderness. Our literature supports the idea that each of these performs an important role in our psychic economy, and that quite apart from nostalgia, sentiment, or any narrow measures of utility, either economic or recreational, each offers indispensable satisfactions. Hence the prospect of the disappearance of any one of them, or of the irrevocable blurring of the boundaries between them, as in the spread of suburbia, would be an intolerable loss. The literary pastoral emphasizes the value of contrast as a mental resource, and supports the views of those ecologists who define the relation between the urban and extra-urban environments as a form of symbiosis.

Our pastoralism would therefore seem to confirm the opinion, advanced by Paul Goodman and others, that our society requires rural as well as urban reconstruction. Indeed, urban renewal without rural renewal is self-defeating, if only because rural decay is driving an impossibly large population into our cities. To make our small towns and vast countryside economically viable and culturally interesting is an indispensable aspect of solving the urban problem. Besides, the countryside should be made available to our city dwellers, and not merely a select few, as a necessary retreat from the nerve-racking demands of our complex civilization. Here again we see the relevance of the symbiotic relation that the pastoral design tacitly establishes between the contrasting environments. In Western society the wealthy and aristocratic always have appreciated the advantages of periodic retreats from the world, of moral and physical holidays from complexity. Today, in the United States, when only a small fraction of the population can be classified as rural, enjoyment of the land itself is denied to most people. What is needed is not the extension of suburbia or the proliferation of commercial resorts, but the invention of means whereby city dwellers can temporarily enjoy the pleasures of an alternative way of life. To make such facilities available to a much larger segment of our population would seem a legitimate goal for planners in an affluent, democratic society.

My final suggestion brings us back to the realistic implications of the third stage of the pastoral design: the return. Our literary pastoralists, surprisingly enough, reinforce the inescapable lesson of common sense, namely, that the mainstream of contemporary history is to be found in the urban-industrial environment. If many people feel the urge to retreat, it is in some measure an effort to invest their lives with a sense of order and meaning that is lacking in the world of the machine. The curious tendency to find more significance in the seemingly haphazard dispersal of trees, animals, and hills than in the relatively deliberate patterning of streets, buildings, and parks, cannot be wholly attributed to the actual physical character of the two settings. The flight to suburbia, which might after all be described as a debased and doomed version of the pastoral retreat, is in part at least a gesture of revulsion at the chaos, contradiction, and nonmeaning that we associate with our cities. It is, by the same token, an effort somehow to recapture certain social and political attributes of smaller communities.

In arguing for the importance of the social and political as well as psychological motives for the effort to escape the city, I do not mean to discount the impetus provided by "real" physical discomfort, ethnic prejudices, and sheer ugliness. But there is reason to believe that the widespread preference for the suburban "country" over the city (and however foolish the idea may seem, the commuter on his way to Levittown does say that he is on his way to the "country"), also derives from the symbolic significance of the two settings. That is why it would be useful to learn more about the values that people attach to various forms of pastoral retreat, and to distinguish between those aspects of rural or wilderness living that are capable of fulfillment, and those that are illusory. For it is possible that planners could find ways to provide some of those satisfactions within the city. They could accomplish this purpose both in the traditional manner, by reproducing certain physical conditions of rural life (parks, playgrounds, open spaces), but also by taking into consideration the need for social and political surrogates for rural and small town institutions. Our literary pastoralism suggests that physical planning without political (to use the word in its broadest sense) planning is futile. What I am saying, in short, is that today the planner finds himself in a position analogous to that of the pastoral figure at the conclusion of an American fable. His problem is to find ways of creating, within the urban environment, that sense of belonging to an orderly pattern of life that has for so long been associated with the relatively unspoiled, natural landscape.

Notes

This chapter was one of 12 papers delivered at the Smithsonian Institution Annual Symposium, February 16–18, 1967 and published as the Smithsonian Annual II, *The Fitness of Man's Environment,* with an introduction by Jennie Lee, Smithsonian Institution Press, City of Washington, 1968.

1. Robert Frost, "Letter to William Stanley Brathwaite," in Mark Richardson (ed.), *The Collected Prose of Robert Frost* (Cambridge, MA: Harvard University Press, 2008), xvii.
2. "Robert Frost in conversation with Cecil Day Lewis," *Claremont Quarterly* (Spring) (1958): 13.
3. Ernest Hemingway, "Big Two-Hearted River: Part I," in *Hemingway on Fishing* (New York: Simon and Schuster, 2002), 7.
4. Ibid., 3.
5. Ibid., 9.
6. Ibid., 23.
7. Ibid., 24.
8. Lewis Mumford, "Constancy and Change," *New Yorker* (March 6, 1965), 57.

CHAPTER FIVE

Boys in the City: Homoerotic Desire and the Urban Refuge in Early Twentieth-Century Germany

Elena Mancini

Arguably more than any other aspect of a Western industrialized economy, urbanization has contributed to the social organization of homoerotic life and a freer expression of alternative gender roles and identities.[1] Lack of acceptance and negative sentiments toward queer communities and lifestyles are often found in accompaniment with antipathies toward symbolic and experiential realities of life in the city. Despite the legal obstacles and social inhibitions toward the expression of homoerotic identity that continue to persist throughout many cities of the world, the city continues to provide a refuge, or at a minimum, a tolerant haven of sorts for queer life well through the present.[2]

That higher degrees of tolerance and more progressive stances toward alternative gender identity can be found in urban areas as opposed to rural and suburban areas constitutes a reality that is both empirical and symbolic for the United States and most of the Western world. On an empirical plane, urban life's increased anonymity, the indiscriminate accessibility of public spaces, and the social tolerance demanded by high population densities promote greater possibility of alternative gender expression. On the symbolic level, since the myth of Babylon, the city has been linked with power and bold innovation and their underbelly of excess, moral transgression, and chaos.

The demonization of the urban has been witnessing a vigorous resurgence in contemporary American political and popular culture. Populist politicians of the likes of Sarah Palin have been remarkably successful in leveraging negative stereotypes about urban life to galvanize fears and distrust of the city and thereby earn political favor through false identification with average rural voters. In the rhetoric of such demagogues, urban politicians are painted as arrogant, ungodly, elitist, morally bankrupt, and out of touch with the concerns of ordinary, working citizens. This antiurbanist venom sets up a trumped-up dialectic between urban and rural areas. Accordingly, rural areas are construed as the loci of unfettered patriotism, wholesome morals and traditional family values. An egregious display of this unbridled disparaging of the urban was the late conservative evangelical Jerry Falwell's incendiary claim that the September 11 attacks on the World Trade Center were caused by "gays, feminists, pagans and abortionists."[3]

This concept of the menacing city that has loomed in the collective imaginary of Western culture since the dawn of Old Testament, witnessed an animated return in the Weimar Republic. The widespread perception of the boding, all-engulfing metropolis of Berlin was pervasive and received vivid and often times shrill expression in the art, film, and literature of the German Weimar period.[4] In terms both artistic and rhetorical, Berlin's modern urban expansion was rendered into a vision of the mythic Babylon, both a whore and a veritable locus of a modern Apocalypse. Although the cultural and artistic output of the Weimar Republic offers a rich storehouse of these images, Spengler's metaphors of Western decline, Döblin's *Alexanderplatz* and Lang's expressionist filmic oeuvre offer among the most powerful representations of the city's tantalizing ominousness.

The breakdown of traditional gender roles and the greater visibility of alternative gender expression became synonymous with the urban metropolises like Berlin in late Imperial Germany and the Weimar Republic. The well-known factors such as World War I, the rise of industrial economy and women in the workplace, urban population surges, the rise of a culture of mass entertainment and the topography of the city contributed to a rapid transformation of gender codes.[5] Urban industrialization and World War I were among the principle factors contributing to the masculinization of women and the feminization of men. The need for women in the workplace that emerged with men at the war front had a profound effect on the power dynamics between men and women. Women gained in power, confidence, and independence and men felt themselves weakened, dispensable, and emasculated.

The image of the pant-wearing, cropped hairstyle "new woman" was emblematic of these gender transformations.

Because the economic conditions for the employment of women were most favorable in cities, it only followed that the relaxing of rigid gender norms would be strongest there. Therefore urban life attracted not only those who sought economic opportunities, but those who did not identify with normative, heterosexual ideals of gender identity as well. At the same time, the challenge to traditional gender roles also generated anxiety about the disintegration of the moral order and the cultural and religious mores that composed a sense of unified national identity.[6] For instance, the ideal of German motherhood was elevated to a paradigmatic expression of German femininity and a point of national pride.[7] This chapter will argue that the angst toward the modern societal transformations described above often coalesced with antiurbanist sentiment. It will do so by examining how this angst toward the gender revolutions that accompanied urbanity transpired in the literature of the period and reveal similarities with present-day rhetoric.

The city as a cultural and economic locus played a critical role in the organization of alternative—and expanded—sexual expressions and relations among the genders in Imperial Germany and the Weimar Republic.[8] The rhetoric and anxious tirades of cultural and religious conservatives throughout nineteenth and twentieth centuries betray anxieties about city life that are tantamount to branding them as hotbeds for transgressive behavior and for enabling modes of social interaction that catalyzed the dissolution of wholesome family life. This chapter will explore urban development and its impact on the representation of homoerotic relationships and affectional ties outside of the nuclear family in the early twentieth-century German context. It will specifically do so by examining three novels of the period featuring queer protagonists and their struggles to affirm themselves outside of the emotional and legal confines of the traditional nuclear family.

Urban sprawl was greeted with a great deal of ambivalence on the part of pastors, clergymen, and social reformers in Imperial Germany. The explosive increase in urban population and the freedom and opportunity embedded in city life were seen as developments that threatened to undermine the fabric of family life embedded in traditional Christian values. To help put this rate of growth in numerical perspective, the city of Berlin registered a population increase of roughly 250 percent from 1871 to 1910. The responses to this phenomenon were varied and yielded positions that were both extreme and often contradictory on the part of cultural conservatives. I will point to two basic and distinct approaches that were espoused by a generally conservative outlook toward the question of urban expansion.[9]

The two camps in which conservatives were divided were the antiurbanists and the social reformers. Antiurbanists fought to curtail urban expansion by demonizing city life and depicting it as neurotic and morally decadent. Many of the ideas taken up by the antiurbanists were enshrined in Max Nordau's 1895 opus *Degeneration*, which was essentially a no-holds-barred indictment of modernity, urban culture, and the cultural decadence of the fin de siècle period. The social reformers' camp on the other hand pursued their concerns toward the preservation of cultural and religious values, by instituting social programs and infrastructures to improve the quality of life in the city.

A critical analysis of the attitudes and assumptions that these novels present toward the city will reveal the structural, cultural, and psychological reach of homophobia and illuminate the sociology of homoerotic desire. This chapter will engage the notion that cities not only tolerate but facilitate and possibly even promote homoerotic attachments and alternative affectional bonds outside of traditional nuclear families. In the early 1990s, gender historian John D'Emilio[10] argued that since the emergence of modern capitalism, cities have acted both as the breeding grounds and the supporting structures for affectional relations outside of the emotional and legal confines of the nuclear family in myriad ways. Both psychically and concretely, modern cities have had a liberating effect on subjects who sought modes of living and loving in social arrangements that transcended the confines of the nuclear family. The vast array of social networks and mechanisms with which modern cities were equipped liberated individuals from the rituals, obligations, and emotional constrictions associated with family life and afforded them the freedom to create alternative types of families or communities bound by affection rather than blood relations. D'Emilio termed these ideal types of communities "affectional communities"[11] and made a strong case as to why these should enjoy the same rights and legitimation as the nuclear family.

Antiurbanist prejudices often dovetailed with homophobic discourse. While it can rightly be argued that the affectional bonds I am describing here have also existed outside of the organizing structure of the city—historically, formal institutions such the military, prisons, boarding schools as well as sex-segregated religious communities and acting troupes have also fostered affectional relationships that are not bound by the marital seal. A distinction worth noting between the bonds formed in these settings and those that occurred in metropolitan areas was that happenstance and impermanence seemed to play a more decisive role in the former, whereas the relationships that arose in the city were more deliberate. This was largely because both structurally and sociologically, the city offered the economy, infrastructure, public space, and as a consequence mentality and demographic that enabled alternative modes of sexual interaction and expression. Individuals aware of these factors could consciously choose to have an identity or experience outside of the legal and sanctioned sexual norm.

Critical empirical insights into the socialization of homosexuals and other sexual minorities of the period (lesbians, transsexuals, transgenders, hustlers, prostitutes, and hermaphrodites) in the city of Berlin were offered by homosexual rights activist and sexologist, Magnus Hirschfeld. Myriad unprecedented and eye-opening descriptions of Berlin's motley homosexual scene appear in his, *Das dritte Geschlecht*[12] (The Third Sex) an ethnographic study of Berlin's homosexuals published in 1904. Here Hirschfeld painted a detailed portrait of the goings-on in the underground cauldron of homoerotic Berlin. According to Hirschfeld, between 1 and 2 percent of Berlin's 2,500,000 inhabitants were homosexual. This segment of the population included transvestites, transgendered individuals, pederasts, and other sexual minorities. From the boisterous, pulsating vitality of Berlin's homosexual bars to the complex psychologies of individuals who harbored their homoerotic longings in seclusion from any community identification or awareness, Hirschfeld reported on the colorful patchwork of indeterminate erotic desire that flourished in Berlin. He drew an intimate connection between the geography of the city and the possibilities it allowed for the explorations of alternative desire. The city of Berlin, he argued, with its multitude of tunnels, train stations, and public baths, was able to support a richly functional and diverse architecture of homoerotic desire. Characterized by large expanses of land from north to south and east to west, this urban metropolis was well poised to furnish the panoply of crucial elements needed to sustain homoerotic affectional bonds in a Germany burdened by Paragraph 175, the infamous antihomosexual statute that remained in German Criminal Code from 1871 through 1968 in East Germany and 1969 in West Germany. Anonymity, legions of hiding places, a communal sense of like-affected individuals, and the possibility to live at great enough distances from the nuclear family without needing to relocate to another city were among the many advantages that Berlin had to offer to same sex relationships. Hirschfeld pointed out a number of cases in which it was possible for native Berliners who were homosexual to continue living in Berlin and not encounter family members for over two decades. The fact that Berlin could provide the voluntary estrangement that many in the homosexual community required made it a veritable safe haven of sorts for homoerotic life at the turn of the twentieth century.

Although the topography of the city and the anonymity it provided were crucial to the formation and maintenance of homoerotic relationships, there are other factors that aided the emergence of these alternative affectional bonds. D'Emilio argued that a free labor market was critical to the emergence of gay and lesbian identities in the early twentieth century and put forth the view that gay and lesbian sexual orientation as an outgrowth of a modern, urban capitalist economy comprised of wage labor in exchange for commodity production. Contrary to the medical theories that had gained wide popularity at the end of the nineteenth century that had argued the biological grounding of homosexuality, D'Emilio argued that while some people may be born with the same sex erotic desire, the establishment of a gay and lesbian group identity was only made possible by socioeconomic conditions of modern capitalism.

> By the second half of the nineteenth century the situation was noticeably changing as the capitalist system of free labor took hold. Only when individuals began to make their living through wage labor, instead of as parts of an interdependent family unit, was it possible for homosexual desire to coalesce into a personal identity—an identity based on the ability to remain outside the heterosexual family and to construct a personal life based on attraction to one's own sex. By the end of the century, a class of men and women existed who recognized their erotic interest in their own sex, saw it as a trait that set them apart from the majority, and sought others like themselves.[13]

D'Emilio attributed much of this social phenomenon to the transition from an agrarian to an industrialized economy, which divested the family of its economic functions and labor production that hinged on familial interdependence. These new economic realities also contributed to large population shifts toward the city.

According to D'Emilio, the social hegemony of the nuclear family in the twentieth century was also a product of modern capitalism, As long as procreation and childrearing was viewed as a privatized enterprise and conducted within the strict boundaries of the nuclear family—alternate forms of affectional bonds and erotic expression would not receive legitimacy.

> Capitalism has led to the separation of sexuality from procreation. Human sexual desire need no longer be harnessed to reproductive imperatives, to procreation; its expression has increasingly entered the realm of choice. Lesbians and homosexuals most clearly embody the potential of this spirit, since our gay relationships stand entirely outside a procreative framework. The acceptance of our erotic choices ultimately depends on the degree to which society is willing to affirm sexual expression as a form of play, positive and life-enhancing.[14]

More recently, Ethan Watters,[15] in his 2003 sociological study of millennial singles, *Urban Tribes: A Generation Redefines Friendship, Family and Commitment*, has attempted to provide a narrative[16] for the current generation of individuals who choose to spend long periods of time outside of traditional family units, also sees the city as central to these alternative social formations, which he aptly calls "urban tribes." Watters, too, highlights the role of the city in creating vast networks of friendships and nonbiological families in sustaining individuals socially, emotionally, psychologically, and in some cases economically as well. From hosting holiday gatherings to bailing each other out of financial jams, Watters delineates the myriad of ways urban tribes have taken over many of the traditional functions of the family.

One century earlier, Hirschfeld had argued the nonpathological nature of homosexuality and had advocated the nonconflation of procreation and love.

> As soon as people realize that reproduction is not the exclusive goal of love, the phenomenon of homosexuality, so enigmatic under its assumption, loses much of its puzzling nature, and to a still larger extent, when people admit that love is also productive whenever it does not issue any new creature, that a spiritual procreation also exists, and that the value of persons depends on the values they create, no matter if the creations are of a material or spiritual kind. If love principally serves to enhance one's own happiness and that of others, then it is incomprehensible why it should not also extend to include persons of the same sex.[17]

Not only did Hirschfeld recognize that the formation of alternative forms of affectional bonds and communities were crucially beneficial for individuals[18] who could not embrace traditional sexual mores and social arrangements, but he also saw that the city of Berlin offered the unique sort of tolerance, locales, and topographical infrastructure to enable these communities. As a clinician and therapist of sorts for homosexuals and everyone who fell outside of the legal sexual norm, Hirschfeld also actively facilitated the creation of alternative communities by hosting public lectures and events specifically centered on non-normative sexuality. He was engaged in expanding homosexual social networks and encouraged reclusive, closeted homosexuals to frequent the therapy groups that he hosted at his Institute for Sexual Studies and to attend the homosexual bars and transvestite balls that Berlin boasted.

It was precisely the recognition of this liberationist spirit of the city that fomented an avid antiurbanist sentiment in the German fin de siécle. Andrew Lees's urban history of Imperial Germany[19] reveals the character and extent of culturally and politically loaded biases against the city. Lees shows the broad range of social ills and moral failings that the antiurbanists linked to the rise of city. Crime (prostitution and alcoholism), moral laxity, the rabid decline of religious observance, deterioration of family life, the disregard of aesthetics and growing political valence of socialism, and how these could ultimately engender a cultural and racial degeneration were among the fears most frequently cited by antiurbanists. The competition of urban life was identified as a primary source for moral decline.

> The big city was a school for pushiness [Strebertum]. Protestant clergyman Rogge claimed that the big city was "the dwelling place for masses of criminals. Not only mass misery but also mass degeneration pervades it. Any army of prostitutes and pimps eats away at its foundations."[20]

Often fear of the city overlapped with racial and nationalistic anxieties. The growth of cities brought with it a concomitant fear of a menacing population decline. Since the city was viewed as a cause for a weakening of the nation, and a potential cause for the decrease in size and effectiveness of the German armed forces, this led to an invocation of xenophobic anxieties. A writer quoted in Lees remarks: "When there is no more German migration, then there are Poles, Czechs, Gypsies, Mongols.... In this way, men restlessly build a Tower of Babel, so that one day linguistic chaos will reign."[21] Classen admonished that "the growing urban presence of Italians and especially of Poles confronted Germany with the danger of racial degeneration."[22] Other indictments against the city were politically inflected. Wapler cautioned that "socialist agitation took root in the big city and spread from there to the countryside."[23] Rogge argued that "in politically turbulent times, big cities are hotbeds of revolutions."[24]

In this chapter I will demonstrate how the cities have served as critical loci for sustaining nonmarital affections of the queer kind and examine this phenomenon within the framework of homosexual discourse at the turn of the twentieth century by considering three novels of the Imperial and Weimar period. My thesis is that the antiurbanist discourse embedded in these texts mediates the predominant attitudes regarding homosexuality in the early twentieth century.

For this study, I have chosen to examine Klaus Mann's 1925 novel, *Der fromme Tanz* (The Pious Dance), John Henry Mackay's 1926 novel, *Der Puppenjunge* (The Hustler: The Story of a Nameless Love from Friedrich Street) that Mackay published under the pseudonym of Sagitta[25] and Stefan Zweig's 1929 novel, *Die Verwirrung der Gefühle* (The Confusion of Feelings). I propose viewing the city as it transpires in these texts as social and cultural matrices of homoerotic and transgressive desire. I will argue how positive and negative depictions of city life uncover varying types of prejudicial attitudes toward homoerotic love. More specifically, a close analysis of the treatment of the city in these novels will not only corroborate many of Magnus Hirschfeld's observations on homoeroticism and the city, but will also furnish a less theoretical and more affective dimension to this problematic. It will be demonstrated how the antiurbanist discourse embedded in these portrayals of the homoeroticism, in varying degrees, frustrate the emancipatory potential for alternative gender expression that the city could yield. In each of these novels, it is as if the deviation from traditional script of masculinity and normative sexual expression that the protagonists undergo is so stark that a return to precity life is conjured as a redemptive solution for each protagonist.

I have selected these specific novels because they offer penetrating insights into the sociopolitical dynamics of male homoerotic desire during the interwar period. Broadly, all three of these novels can be read as a queer *Bildungsroman*. In each novel, the main protagonist experiences the physical transition from the provincial bourgeois family into the secular urban setting as a launching into a sexual vortex. Individually, each of these novels offers a unique and compelling insights into the sociodynamics of early twentieth-century queer desire. Viewed in relation with one another, these novels weave a tapestry of homoerotic desire that is heterogeneous and complex.

In Klaus Mann's *Der fromme Tanz*, the aspiring artist Andreas Magnus leaves his hometown and staid bourgeois family house and for Berlin in hopes of pursuing his artistic ambitions. Andreas, identifying with the lost generation that lacked the purpose and discipline of the previous one that had served in World War I, and without artistic talent to speak of, abandons his quest for artistic development in favor of indiscriminate human experiences. In the city, Andreas is immediately confronted with greed, harshness, and financial stress. Monkeyed by his first host family, Andreas moves into a boarding house managed by a business savvy, pseudoaristocratic, Vaudevillian type performer, Fräulein Franziska. He agrees to perform show ballads with her on stage in order to pay the rent. In this milieu, the world of wanton excess and sexual debauchery opens up to Andreas. Prostitution, cocaine use, unwanted pregnancies, and transvestitism are the themes that mark the lives of those who are part of his every day. The son of a well-to-do bourgeois artist, Andreas experienced an immediate revulsion toward Berlin and the harshness and moral dissoluteness that seemed endemic to it.

> Berlin was big. Although Andreas hated it with every fiber of his being, had indeed abhorred it from the very morning it had afflicted him with its merciless ugliness, quite like a nightmare vision—nevertheless every day and every night he wanted to circulate in it again, full of reverence and humility, in its unfathomable, mysterious, inexhaustible magnitude.[26]

By contrast the vision of a hill in the province that overlooks the city in the protagonist's dream represents pristine, unadulterated safeness.

> The place must have smelled of incense and closeness, he now noticed for the first time. Outside the air was so pure and clear. Apparently he was on a hill overlooking a big city. But he did not recognize he city, for it was blurred and formless as water at his feet on a dark night. Behind him, a little higher up, a white church gleamed, arching loftily in the night.—Andreas walked a little farther, holding the rosary. The white road ran before him like a purling stream, downhill towards the city that roared and buzzed in the distance... What miracle was inconceivable the wanderer asked himself—above this city?[27]

The possibility for miracles or well-being for the protagonist resides only above or outside of the city.

Unremittingly in love with a vagrant artist type, Niells, the protagonist is led to move to both Hamburg and Paris in pursuit of his beloved. Andreas's devotion toward the decidedly egotistical and emotionally shallow Niells manifests itself as a spiritual and artistic vocation. Andreas loves Niells with selfless and idealistic abandon. He sees his feelings as morally ennobling and is prepared to suffer or even die for his love. Reading religious significance into his suffering, he sees this love as a passionate way of the cross and in the spiritual lineage of Herman Bang, Paul Verlaine, and Oscar Wilde. As can easily be anticipated, Andreas's love remains unrequited. While it is unclear whether he returns to his family of mundane artists, who in the interim have become a source for moral support for him or his biological family. However, psychically, he does return to the province by way of dream. Andreas redreams the dream with which the novel opened and that is of the Blessed Mother not accepting his offering of rosary beads because he had not yet suffered enough to understand the meaning of offering. In the subsequent version of the dream, Andreas came to understand the suffering that the Blessed Mother spoke of and offers the rosary beads to his beloved by wrapping them around a picture of him. This sublimated metaphor and others like these throughout the novel support a commonly held view of homoerotic love in the early twentieth century and that is that it should be experienced as an exercise in spiritual edification through self-denial.

In *The Hustler,*[28] the orphaned adolescent protagonist Gunther Niells also flees the nauseating quaintness of the province in search of inspiration and adventure. He decides to run away from the home of his care-giving aunt and uncle in order to experience for himself the myth and fascination of Berlin that his friend Max Friedrichsen back home described to him in the most vibrant of metaphors. Guileless and without any real friends in Berlin, Gunther's financial vulnerability quickly exposes him to the ruthless clutches of Atze—a male hustler-pimp, prompting Gunther's rapid descent into Berlin's underworld of male hustlers. The novel foregrounds the topography of Berlin and its advantages for the practice of illicit desire. Sites like the Stettin tunnel, the passageways, Friedrichstrasse, and the Tauetzien all figure as choice urban refuges for the commodification of male homosexual sex. Gunther or "Chick" as he will later become known on the streets, quickly becomes initiated in the vast panoply of bars and nightclubs frequented by hustlers, homosexuals, transvestites, and sexual outsiders, and assumes the ranks of Berlin's fast-talking, street youths shortly after his arrival to the big city.

Although risky and irregular, hustling in Berlin proves to be an effective way of eking out a living for Gunther and the other male hookers. These bars, however, are continually scoured by blackmailers and policemen and represent principal sources of stress outside of money for the street hustlers. The span of Chick's clientele ranges from slovenly proletarians to wealthy counts. Chick hustles without harboring any moral quandary. For him it is merely a job. Detachedly, he describes some of the sexual acts as nothing more than what the parish pastor had done to him back home. He eschews any type of emotional bonds with these men and feels importuned whenever genuine love is proffered to him. This is certainly the case with Hermann Graff, a middle-class publishing agent who exhibits a deep and sacrificial love for him.

Graff adores Gunther and sees him as his soul mate. He harbors an overly romanticized image of Gunther and believes him to be purehearted but plagued by misfortune all his life and thereby denied a decent bourgeois existence. He refuses to even consider that hustling might be Chick's occupation of choice. Graf pursues Gunther as a friend and as a potential partner but Gunther only sees Graf as a john and a reliable meal ticket. When Graf gives Gunter five Marks at the end of every date to help him get by, Gunther sees it as strange that Graf doesn't even ask for sex. He wishes Graf would just ask for sex rather than burden him with probing emotional questions and demands for time. Eventually, Gunther comes to depend on Graf for money. After experiencing illness and extensive abuse on the streets, he begins to form an emotional attachment to Graf and decides to give up hustling. The novel ends tragically with Gunther's being in the wrong place at the wrong time—standing in front of a hustler bar to lend some money to a hustler friend, he gets arrested by the police for offenses against Paragraph 175. Consumed with worry as he waits for Gunther in vain, Graf pays a young man from Gunther's hustling network for any information he can give him on his beloved. The hustler informs him that Gunther is in prison at Neuenhagen.

At the height of his desperation, Graf writes Gunther an emotionally incriminating letter, which the police intercept. As a result, Graf is brought to trial. Gunther is also summoned to court. Now with a hardened, emaciated look and a shaven head, Gunther is barely recognizable to Graf. Both men are charged with committing indecent acts. Banished of all his strengths, Gunther admits to all of the charges.

Graf is arrested and interred for two months in prison. After being released from his incarceration, the count finds kindness where he least expected it. He receives a letter from a long lost aunt who resides outside of Munich. Expressing her sympathy with him, Graf's aunt invites him to come and live with her. Graf leaves Berlin to visit her and is succored by her understanding. Having been many years in a platonic marriage with a man who also desired in the way that Graf did, his aunt empathized with and witnessed firsthand the suffering and denial that accompanied the love that dared not speak its name. She genuinely wished to offer her friendship and solidarity to Graf and others who suffered his fate.

What was novel and remarkable about *The Hustler* was not only that it depicted homoerotic life as an integral part of life in the city, and not a mere fringe phenomenon, but that it also vividly underscored how economic realities impinged upon homoerotic interaction. Men seeking same sex fulfillment seek men who offer it for economic compensation. The pace of city life and its economic contingencies seem to preclude deep, meaningful homoerotic bonds. Quickies and pseudo-intimate encounters are the modes of sexual expression that best seem to suit city life. This is also due in large part to the criminalization of homosexuality in Wilhelmine Germany and the harshening of the penalties against it by 1929.[29] The risks of being caught and persecuted were palpable and real in the streets of Weimar Berlin. These realities leave Hermann Graf with his longing for a meaningful spiritual union feeling alienated and displaced in Berlin. In spite of these frustrated hopes and desires, the novel presents us with a Berlin that features a broad spectrum of same sex loving individuals. From economically motivated youths to effete same sex desiring counts, the novel shows how the city could embrace and accommodate all of these varied desires. Similar to Hirschfeld's *Berlin's Third Sex*, the novel shows how the Berlin bar scene catered to the diverse palette of sexual others. It also depicts the social problems and the inherent threats and limitations to pursuing alternative modes of desire.

Stefan Zweig's *Verwirrung der Gefühle*[30] presents the city and the province in stark, unequivocal contrasts. While most of the novel unfolds in the province and the descriptions of the city seem mostly peripheral and abstract, the distinctions that are drawn are incisive. This polarization not only aligns with traditional assumptions about erotic life in the big city but also reveal their embeddedness in narrow typologies of homoerotic love. The novel opens with a narrative flashback: the main protagonist, whom the reader knows only as Roland, an accomplished philologist in the English letters, indulges in a sentimental recollection of his youth sparked by the tribute his students pay to him in honor of his sixtieth birthday. Roland experiences a feeling of deep unease upon hearing the speeches in his honor. The life that he hears described—accurate down to the last scholarly publication—does not feel like his own. It feels cold and impersonal as it relates neither to his emotional life nor to the real source of his passion for learning. This alienation sets the stage for Roland's first person narrative of his troubled and dissolute youth.

Brought up in a bourgeois family from a small town, Roland is requisitely sent to university to Berlin to study classical philology. Somewhat surprisingly he manages to persuade his resolutely conservative and domineering family to allow him to study English philology instead. However, rather than pursuing his studies with the zeal and commitment that were expected of him, Roland surrenders himself to the titillating world of copious drink and prodigious erotic adventures that Berlin had to offer him. Over time, his life in the big city becomes debauched to the point of altogether neglect university life. A surprise visit from his father while engaging in a sexual tryst with a young woman causes Roland not only to attract paternal scorn but to abandon his studies in Berlin definitively. Roland's father insists that it lies in his son's best interest to attend university in a provincial town where decadent distractions were notably lacking. In the unnamed small university town, a renowned Shakespearean philologist takes the young collegiate under his wing. The professor awakens in Roland not only a vibrant zeal for acquiring knowledge but the desire to cultivate a deep affectional bond with him. The professor confounds and disquiets the young man by vacillating between the extremes of reciprocating Roland's desire to displaying an attitude of complete neglect toward Roland. Married and unable to freely surrender to his homoerotic passions, the professor finds his only sexual outlet in regular excursions to a nearby urban center where he frequents disreputable bars and surrounds himself to the company of male prostitutes. Roland, on his part, weary of experiencing his desires for the professor continually rebuffed, decides to triangulate his feelings for him by initiating a relationship with the professor's wife, who is sexually vulnerable as her marriage to the professor is platonic. It is both the desire for revenge toward the professor and a sense of identification with his wife that Roland undertakes a sexual relationship with her. Satisfaction quickly subsides and Roland finds himself plagued with guilt for having betrayed his beloved mentor. This drives him to leave the professor's house on amicable terms. His leave-taking of the professor prompts the most intimate and surprising of revelations. The professor declares an exalted spiritual love for Roland—one that not only inspired him but awakened the most deeply buried erotic longings in him. He tells Roland of his love for young men and the humiliations, disappointments, and renunciations to happiness that he has had to endure because of it. This confession leads to his recounting an episode from his youth in Berlin. His youthful submission to his emotional yearnings subjected him to homosexual blackmail but nearly jeopardized his position at the university. For this reason, he explains, he is resigned to depriving himself of a fulfilling love in order to privilege his intellectual and spiritual vocations and to draw inspiration from the youth, which surrounds him. The leave-taking culminates with the professor emotionally stirred and kissing Roland passionately for the first and last time. The novel ends with the older Roland (40 years later in flash forward—by which point he is married with children) remembering this kiss and declaring that he has never loved anyone as much as he loved this professor.

Verwirrung der Gefühle makes significant inroads into the principal debates on homosexuality in the early part of the twentieth century. Its treatment of the city offers a prismic view of the constrained realms of possibility for male homoerotic expression. It foregrounds the homosexual archetypes that these debates engendered at the height of the homosexual liberation movement at the turn of the twentieth century. For this purpose it is most instructive to focus on and unpack some of the tensions and oppositions that come forth in this novel and to analyze the symbolic weight yielded by these. In ways that are not at all subtle, the city is pitted against the province and by extension commodified licentious homosexual carnality versus spiritual and chaste pedagogical eros. These two extremes are tethered to the most basic conceptualizations for male homoerotic expression at the turn of the twentieth century.

Early in the novella, Roland experiences Berlin as both a site of electrifying erotic exuberance and as a locus for moral degradation. For him, the city of Berlin captures both masculine and feminine principles. He anthropomorphizes he city. At first, Berlin seems, both topographically and technologically, masculine to him. The stones, the electricity of the streets, the hot-tempered, pulsating tempo and the hungry greed of the city evoked an inherent masculinity to him (my translation, p. 229). Far away from the orderly, Protestant petite bourgeoisie of his childhood—he experiences a symbiosis with the city—both he and the city were boys who were out to explore the world—both vibrating like a dynamo of restlessness and impatience He later goes on to equate this same city with the impatience and power of a giant woman (Riesenweib) (my translation, p. 229).

This description also reveals a vital and porous interplay between the city and desire itself similar to what Walter Benjamin noted in his reflections on café-culture in *A Berlin Chronicle*. "The time had not yet arrived when the frequenting of cafés was a daily need, and it can hardly have been Berlin that fostered this vice in me, however well the vice later adapted itself to the establishments of that city . . ."[31] The vice is not inherently ingrained into the city but it is the city nevertheless that draws out and continues to give sustenance to the vice.

In subsequent passages in the novel, the city is not only figured as a gendered entity but also as having a valence that is sinisterly erotic. The city is depicted as the keeper of the forbidden and dangerous of secrets and is the place to which the professor escapes to find temporary relief for his erotic leanings. It allows him to carry on with his double life. It is also the place in which he discovered the type of erotic expression that seemed most natural and desirable to him. Yet, the professor characterizes Berlin and the homosexual scene that he witnessed there as decisively abhorrent and morally abject. He describes an underworld full of ominous contrasts—shadows and lights, bridges and tunnels, a place inhabited by the vilest of characters eager to exploit one's erotic disposition through blackmail. This underworld was also depicted as shrill and boisterous and inhabited by members of the lowest social classes. Many of these dives the professor describes featured the stylized male hustlers, perfumed barbers' assistants, excited giggles of transvestites, and wandering, money-hungry actors.

It is thus clear that *Verwirrung der Gefühle* posits a clear hierarchy of homoerotic desire. This hierarchy is organized around the ideological split of body and intellect/soul. The professor shuns a life that would allow him to embrace homoerotic sexual gratification in order to idealize and, to a certain extent, pursue a spiritual bond with the objects of his desire. While it is clear that he as a homoerotic subject also desires carnally, he chooses a life that allows him to keep his status and pursue an extremely idealized form of his love for young men by educating them and to allow them to function as his muses.

This problematic was central to the competing strands of the homosexual emancipation movements started in the last decade of the nineteenth century. The homosexual aesthetic community founded by Adolf Brand in 1896, the Community of the Self-Owned practiced and prescribed a narrow type of male homoerotic love. Hearkening back to the ideals of male homoeroticism described in Greek classical texts and revivified in Winckelmann's aesthetic maxim of "noble simplicity and quiet grandeur,"[32] the Community of the Self-Owned advocated pederastic relationships between men that harbored not only spiritual and pedagogical connotations but also the absence of carnal expression. This movement vehemently rejected the work of the Magnus Hirschfeld's Scientific Humanitarian Committee (an organization that he and fellow activists founded in 1897 to combat Paragraph 175) and its promotion of tolerance of a broad spectrum of homosexuals, transvestites and other sexual minorities unequivocally. The so-called Tunten or Weiblinge fairies or sissies that Hirschfeld routinely examined were categorically abhorred by this group as elements of national degeneration. This disavowal of the effeminate is not only indicative of a chauvinistic brand of masculinity but is intimately tied to late nineteenth century anti-Semitic sexual discourse. As Sander Gilman illuminates in his extensive cultural work on Jewish masculinity, there was a general tendency to regard the Jewish male as having a truncated masculinity. Gilman attributes this biased view of the Jewish male's masculinity to the Christian European synecdochic treatment of the Jewish male to the circumcised penis.[33]

The biological theories that Hirschfeld used to advocate tolerance and understanding toward alternative expressions of gender and sexuality were met with scorn and indifference in this nationalist cultural elitist strand of the homosexual movement. The Community of the Self-Owned refused to accept a type of homosexuality that they deemed anathema to Aryan ideals of masculinity and to the German national spirit. They envisioned male same sex eroticism as an invocation of classical Greece and the great cultural promise that Germany could resurrect. They reviled Hirschfeld's sexual deviants and subterranean bars and sex-affirming attitudes that these members of the third sex[34] seemed to embody.

Zweig's use of the city unlocks some of these debates. The relationship between the professor and Roland seems to follow the erotic paradigms set forth by the program of the Community of the Self-Owned on a number of levels. The type of relationship between the two seems to reflect what late nineteenth-century youth leader and pedagogue, Gustav Wynecken termed, pedagogical eros. The age difference between the professor and Roland is a generational one and the intellectual stimulation and inspiration that the two draw from one another is mutual. Desire for the opposite sex is almost entirely absent, in both of these men and when it is practiced it is motivated exclusively by a wish for immediate sexual gratification of and fulfillment of bourgeois expectations.

In his descriptions of the professor Roland continually likens his semblance to Roman statues and even Socrates. The bourgeois values of self-restraint and respectability as outlined in George L. Mosse's landmark historical work[35] on bourgeois masculinity at the end of the previous century are reflected here. The novel plays out the classic dichotomy of body and intellect and extends them to city and province. Bodily pleasures are not only secondary to those of the mind but also have their natural habitat in the city. The city is a well spring of sexual alterity, moral baseness, and sheer indulgence. By sharp contrast, life in the province restores the dominion of the mind over the body.

The novel also seems to gesture toward the antiurbanist sentiments raised by Lees, by projecting high moral and aesthetic values onto the province. Yet, *Verwirrung der Gefühle* falls short of making a blanket value judgment on the city. The city still represents a place of yearning and crucial excursion for the professor. In his depiction of the city, the narrator shows the professor as needing the city in order to give space to his carnal desires. Without sermonizing in any way, the narrator also points to deep-seated social problems in cities in interwar Germany. These were not just limited to an inherent moral inadequacy or spiritual dearth that resided in the city, but rather to the inadequacy of the presiding laws. The professor avoided the city for long periods of time because of the rampant extortion of homosexuals and the legal culpability that charges of homosexuality brought with them. Thanks to Paragraph 175, the homosexual at the turn of the twentieth century was a pariah in a dual sense. He was vulnerable both to extortion through private citizens and the law.

In all three of the novels discussed here, the city serves as a psychic landscape on to which erotic desires are projected, explored, and pursued. In all three novels, the city yields potential for the creation of nonnuclear family affectional community that D'Emilio advocates of and of the urban tribe that Watters sees concentrated in contemporary cities. However, these novels also show the ways in which the law, nationalist culture, social prejudices, and medical science invariably impinge on these groups and endanger their very existence. Nevertheless, the very fact that these types of alternative families could establish themselves at all at the turn of the twentieth century is doubtlessly owed to the social, economic, and geographical infrastructure of the city and in no small part to the mystique of city life, of which Walter Benjamin was keenly aware.

Notes

1. For an interesting discussion on modernity and homoerotic life, see John D'Emilio, "Capitalism and Gay Identity," in Larry Gross and James D. Woods (eds.), *The Columbia Reader: On Lesbians and Gay Men in Media, Society and Politics* (New York: Columbia University Press, 1999), 48–55.
2. For discussions on the how the cities support gay life see George Chauncey, *Gay New York: Gender, Urban Culture, and the Making of the Gay Male World, 1890–1940* (New York: Basic Books, 1994); Charles Kaiser, *The Gay Metropolis: The Landmark History of Gay Life in America* (New York: Grove Press, 2007) and Michael Warner, *The Trouble with Normal: Sex, Politics and the Ethics of Queer Life* (Cambridge, MA: Harvard University Press, 1999).
3. For an interesting discussion of the urbanism and Western values, see Ian Buruma, *Occidentalism: The West in the Eyes of Its Enemies* (New York: Penguin Group, 2004).
4. See *The Weimar Republic Sourcebook,* Anton Kaes, Martin Jay, and Edward Dimendberg (eds.) (California: University of California Press, 1994) for a comprehensive guide to the art, history, and politics of the Weimar Republic.
5. For a great overview of the cultural, political transformations ushered by the Weimar Republic, see Peter Gay's landmark work, *Weimar Culture: The Outsider as Insider* (New York: W.W Norton, 2001. 1968). For an excellent varied artistic and philosophical perspectives on mass culture in Weimar Germany, see Siegfried Kracauer, *The Mass Ornament: Weimar Essays,* Thomas Y. Levin (trans.) (Cambridge, MA: Harvard University Press, 1995). For a more detailed focus on the sexual transformations that occurred in this period and tightly drawn connections to Berlin, see: Dorothy Rowe, *Representing Berlin: Sexuality and the City in Imperial and Weimar Germany* (Vermont: Ashgate, 2003).
6. See George L. Mosse, *Nationalism and Sexuality: Respectability and Abnormal Sexuality in Modern Europe* (New York: Howard Fertig, 1985).
7. Mosse described the spectrum of German ideal femininity from Imperial Germany through the Third Reich as ranging from the cult of the chaste Marianne to the domestic goddess of the Nazi era.
8. For different perspectives that illuminate this phenomenon, see Magnus Hirschfeld, *The Homosexuality of Men and Women.* Michael A. Lombardi-Nash (trans.) (New York: Prometheus Books, 2000) and D'Emilio, "Capitalism and Gay Identity."
9. For a comprehensive study of the debates concerning urban development in Imperial Germany, see Andrew Lees's *Cities, Sin, and Social Reform in Imperial Germany* (Ann Arbor: University of Michigan Press, 2002).
10. D'Emilio, "Capitalism and Gay Identity."
11. Ibid., 1.
12. Magnus Hirschfeld, *Berlins Drittes Geschlecht: Schwule und Lesben um 1900,* Manfred Herzer (ed.) (Berlin: Verlag Rosa Winkel, 1904, 1991).
13. D'Emilio, "Capitalism and Gay Identity." 51.
14. Ibid., 53.
15. Ethan Watters, *"Urban" Tribes: A Generation Redefines Friendship, Family and Commitment* (New York: Bloomsbury, 2003).
16. Watters argues the importance for each generation to tell its own story in order to avert having one imposed upon them and that "to live without a story is to live without a sense of coherence and momentum," 10.
17. Magnus Hirschfeld, *The Homosexuality of Men and Women,* Michael A. Lombardi-Nash (trans.) (New York: Prometheus Books, 2000), 369.
18. Hirschfeld appealed for healthy, open, and socially responsible expression of homosexuality.
19. Andrew Lees, *Cities Sin and Social Reform in Imperial Germany* (Ann Arbor: University of Michigan Press, 2002).

20. Ibid., 34.
21. Ibid., 36.
22. Ibid.
23. Ibid., 37.
24. Ibid.
25. Mackay was an anarchist and follower of Max Stirner. Puppenjunge's first printing was only two thousand.
26. Klaus Mann, *The Pious Dance: The Adventure Story of a Young Man*, Laurence Senelick (trans.) (New York: Paj, 1987), 69.
27. Ibid., 7.
28. John Henry Mackay, *The Hustler: The Story of a Nameless Love from Friedrich Street,* Hubert Kennedy (trans.) (Boston: Alyson, 1985).
29. For more on Paragraph 175, see James Steakley, *The Homosexual Emancipation Movement in Germany* (New York: Beaufort Books, 1977, 1993).
30. Stefan Zweig, *Die Verwirrung der Gefühle* (Meisternovellen, Frankfurt am Main: Fischer Taschenbuch Verlag GmbH, 2001).
31. Walter Benjamin, "A Berlin Chronicle," in Peter Demetz (ed.), Edmund Jephcott (trans.) *Reflections: Essays, Aphorisms, Autobiographical Writings* (New York: Schocken Books, 1978) 21.
32. Winckelmann elaborated on his principle of "edele Einfalt und stille Groesse" in his 1755 cornerstone work *Gedanken über die Nachahmung der griechischen Werke in Malerei und Bildhauerkunst* (Thoughts on the Imitation of Greek Works in Painting and Sculpture).
33. See Sander Gilman, *The Jew's Body* (New York: Routledge, 1991).
34. The term "third sex" was coined by nineteenth-century jurist Karl-Heinrich Ulrichs. He used the term to explain same sex love and claimed that same sex loving men, for instance, belonged to the third sex—they were distinguished for having a woman's soul in a man's body.
35. George L. Mosse, *Nationalism and Sexuality: Respectability and Abnormal Sexuality in Modern Europe* (New York: Howard Fertig, 1985).

CHAPTER SIX

*Antiurbanism, New York, and the Early Twentieth-Century American National Imagination**

ANGELA M. BLAKE

In the spring of 1900, at the dawn of a new century, New York City's public image was about to begin a major refurbishment. Due in part to the architectural transformation that occurred in the city between the turn of the century and the end of World War I, during that 20-year period Americans saw reasons to question the late nineteenth-century image of New York as the prime example of the undesirability and un-Americanness of urban life and culture. Between the turn of the twentieth century and the end of World War I, boosters East and West promoted what they argued comprised the uniquely American characteristics of, respectively, New York City's new buildings and the rocky landscapes of states such as Colorado and Arizona. Establishing each place's status as a definitively American landscape, as judged by their promoters, confirmed a key selling point to their consumers in an era of growing cultural nationalism. The direct visual and metaphorical association made by local pundits, architectural critics, and tourism entrepreneurs between New York's growing cluster of downtown skyscrapers and the dramatic mountainous landscapes of the West proved crucial to New York's "branding" as a prime early twentieth-century tourist destination. This Americanization of New York via the representation and interpretation of landscape and commercial architecture moved forward boosters' efforts, begun in the 1890s, to draw middle-class consumers' attention away from the faces, bodies, and buildings of the Lower East Side. Thus they mounted a challenge to a longstanding antiurban sentiment among middle-class Americans, a sentiment whose "proof" had been big cities such as New York, and in particular the social problems of neighborhoods such as the Lower East Side that so dominated the city's public image.

At the start of the twentieth century, writers, illustrators, and photographers noted the emergence of a "new New York."[1] The "old" New York was the city of "darkness and daylight" portrayed in late nineteenth-century urban guides, a place representative of the most egregious aspects of urban life; the "old" New York also suggested the undemocratic divisions between rich and poor of the "old" world of Europe, rather than the equality, liberty, and opportunity of republican America. By contrast, the "new" city, built roughly between 1900 and the end of World War I, consisted of an increasingly splendid architectural showcase, an international financial center, and the home of the nation's leading industrial and commercial powerbrokers. An emergent American national culture, distinct from—and confident enough to challenge—European culture, took root in New York's public and private institutions. This emergent national culture, based in America's largest city, therefore stood poised to challenge the antiurbanism long prevalent in American thought and politics.

The physical transformation of the city, made possible by new wealth, offered the most powerful evidence of a "new," arguably more American, New York, and served to draw attention away from the city's persistent social problems, potential blemishes on the fair complexion of New York's new face. Numerous construction projects such as the first subway line, additional bridges across the East River, and various new private and public buildings made Manhattan the site of great architectural and engineering innovation, the first phase of the cycle of demolition and reconstruction that would characterize Manhattan in the first half of the twentieth century.[2] "The fair new city lies in the embrace of the old one like the new moon in the old moon's arms," declared literary and cultural critic Randall Blackshaw in *The Century*, in an article sumptuously illustrated by up-and-coming architectural illustrator Jules Guérin.[3] "One might almost fancy that the town had been bombarded by a hostile fleet," he wrote, "such rents and gashes appear everywhere in the solid masonry, ranging from the width of a single building to that of a whole block front," Blackshaw exclaimed.[4] Rather than focusing on the inconveniences caused by this disruption, authors in popular monthly magazines as well as those in the more specialized architectural press focused their attention on what these developments in the built environment, in conjunction with other economic and cultural factors, meant for the city's status and public image. Much of this debate centered on whether or not, and for what reasons, New York City could be regarded as the American metropolis, the nation's de facto capital and representative city. Acquisition of such status would represent a major selling point for the city, another string in the city boosters' bow as they worked to brand New York for business and tourism.

Metropolitan status was far easier to confer on the European capitals to which writers on both sides of the Atlantic frequently compared New York. London and Paris, both the long-established capitals of their nations, were centers of political, economic, and cultural power. Paris, especially, formed the national hub in a heavily centralized system of government that had effectively dictated similarly centralized patterns of transportation and commerce. London, overseeing a country whose industrial and commercial development, was less centralized around the metropolis, nevertheless retained a tight political grip both at home and across the enormous British Empire. In comparison with these metropolitan centers, New York—neither politically nor geographically central—might have appeared decidedly provincial. However, the city's boosters claimed, New York was rapidly acquiring some of the attributes of a national—and international—metropolis.[5] In a period when these two European capitals defined the modern and the cosmopolitan, the competition with Europe was vital to establishing New York's new status. For New York to compete with the European capitals, its chroniclers would have to work hard to establish its metropolitan attributes, preferably those that distinguished the American metropolis from the dominant European models. They would also have to overcome prejudices against the idea that American could, or should, be represented by *any* city.

The debate over New York's metropolitan status rested in a larger context. To solidify New York's image as America's representative metropolis meant convincing Americans—non–New Yorkers—that if they visited New York they could "see America." Establishing New York's American identity would clearly entail focusing on what seemed most American during the first two decades of the twentieth century—the city's newest signature architectural feature, the skyscraper.

The debate over New York's metropolitan potential centered on the search for distinctly American attributes, two of which were the city's commercial character and its demographic heterogeneity.

Leading Progressive-era commentator Herbert Croly, writing in *The Architectural Record*, of which he was coeditor, discussed in detail the ways in which New York City might or might not be considered America's metropolis. Croly set a higher standard than did some others for the acquisition of metropolitan status. While he accorded New York the position of the nation's business and commercial metropolis, he argued that to be truly metropolitan the city should also be the social, artistic, and intellectual center of the country. Furthermore, according to Croly, a metropolis should not only reflect national characteristics but also anticipate new ones and remake the old.

Working against New York's metropolitan potential, Croly argued, were not only older national political tendencies that resisted centralization, but also local conditions that contributed to a sense of fragmentation, such as the lack of consistent, coherent urban planning, too great a degree of heterogeneity, and numerous social divisions—from the "unusual proportion of raw and unapproachable foreigners" to the "set of cliques" that composed New York's higher strata. Croly concluded, however, that "New York is national or nothing," and that the consolidation of the city in 1898 combined with the victory over Spain had produced "an outburst of national feeling." As an example of New York's connection to "national life," Croly cited President Theodore Roosevelt's New York nativity, claiming that it was "difficult to see how just such a combination of disposition, experience, training and ideas could have come to a head in any other city." New York, Croly concluded, while perhaps not yet a metropolis by all measures, was "the most national of American cities."[6]

In his address before the New-York Historical Society celebrating its ninety-ninth anniversary in 1903, the essayist and popular lecturer Hamilton Wright Mabie challenged some commonly held national beliefs about New York, and suggested that both the city's heterogeneity and its bold commercialism made it more, not less, American. Mabie referred to the "tradition" of claiming that "whatever New York is, it is not intellectual, religious, moral, homogeneous, beautiful, or American; and New Yorkers have become so accustomed to this state of the provincial mind that they long ago ceased to deny, to explain, or to apologize."[7] Mabie set about to challenge the notion of New York's supposed "un-American" character by suggesting that what he called the "the spirit of the locality" was its "cosmopolitan" character, a spirit that, he argued, was decidedly American. New York's cosmopolitanism was based on racial diversity and political and religious toleration, Mabie claimed. Echoing what Croly had argued, Mabie suggested that the city's heterogeneity had, over the years, "given our friends, south and west, the opportunity of saying that New York is the least American of cities because it is the least homogeneous." Mabie argued the opposite.

In challenging the familiar negative assessment of the city's heterogeneity, Mabie contributed to what became an increasingly familiar debate about early twentieth-century New York. Like some other commentators, Mabie argued that the city's heterogeneity constituted its American identity. " '[I]t is fair to ask,' Mabie contended, 'Which is the most distinctively American, the community in which the citizens are all of one blood, or that in which many races combine to create a new race?' . . . [I]f America stands for a different order of society, a new kind of political and social unity, . . . then New York is the most American of cities." Mabie's promotion of racial diversity can only be regarded as wishful thinking or a plea for tolerance rather than a description of how most early twentieth-century New Yorkers felt about race and immigration. At his time of writing, the number of immigrants entering and residing in Manhattan was increasing rapidly, until by 1910 immigrants made up 41 percent of the population of New York City, the majority from Southern and Eastern Europe. By the 1920s, opposition to the pluralism and "melting pot" image of New York had increased considerably, with many of Mabie's class actively opposing further immigration. However, Mabie's address certainly laid claim on the city's behalf to key tenets of American political ideology.[8]

In honoring the city Mabie, like Croly, emphasized New York's commercial character, describing the city as founded by businessmen and first governed by a commercial monopoly, the West India Company. In an era when the older families of the city were making efforts to distinguish themselves from the *nouveau riche* industrialists, whose money was made not inherited, Mabie's version of New York's history as well as his promotion of the nobility of commerce aimed to bridge the gap between new and old money in New York's upper echelons.[9] According to Mabie, commerce was democratic and besides, he suggested, some of the European cities so admired by Americans as centers of art and civility, as perhaps "above" commerce, were themselves founded on the riches brought by trade: "Commerce is a peaceful and increasingly honest substitute for the wholesale thieving of feudal times. . . . it is well to remember that Venice . . . was the first commercial city of a great period; that her palaces were built because the ships that lay at their doors were laden with the treasures of the East; . . ."[10] Business and trade, these authors argued, *were* American.

New York's profits were not ill-gotten gains such as those acquired by European monarchs and aristocrats; wealth from trade was the product of honest labor, fundamental to the increasing international power of the country. As the undisputed center of the nation's business and trade, New York's metropolitan status, in the eyes of her boosters, was thus unassailable.

National corporations and financial institutions underpinned New York's supremacy as the nation's commercial center. Commerce of all sorts thrived in the city, making use of the presence of large banks, insurance companies, the stock exchange, a huge national and international transportation nexus, local manufacturing facilities, and the nation's largest urban population that provided a ready supply of both workers and consumers.[11] This commercial strength also created a demand for buildings in which to house the growing executive and white-collar workforce of the city's non-manufacturing economy. The limited space of Manhattan Island—especially in the lower portion of the city, traditionally the commercial and financial center of New York—required architects to build on a vertical more than a horizontal plane. The resulting "skyscrapers" multiplied and thrived in New York, particularly in the first two decades of the twentieth century, and made perhaps the greatest contribution to debates about the Americanness of New York City.[12]

The city's burgeoning commercial architecture gave rise to a discussion among architects, critics, and opinion makers of the possibility of a distinctly American architectural style, as well as whether New York might be the center of that development. The terms "skyline" and "skyscraper" were first used in the periodical press in the mid-1890s to describe the new tall office buildings and their effect on the city's horizontal image.[13] Buildings of 10 or more stories made a stark contrast in a horizon still composed of 5- and 6-storied buildings. Previously, the only structures to pierce that veil were the steeples and towers of the city's numerous churches. From 1846 until the last decade of the nineteenth century, the steeple of Trinity church, at 284 feet, was the city's highest point. First eclipsed in 1890 by George B. Post's 309-foot Pulitzer Building on Park Row, the dwarfing of Trinity was, in text and image, a much-remarked on symbol of the city's changing appearance and apparent value system. By 1908 one author could remark, in an article entitled "The City of Dreadful Height," that if Trinity Church was noticeable in the skyline at all it was "conspicuous only as the stub of a broken tooth is conspicuous in a comb."[14]

Despite reservations about the existence of a distinctly American style of architecture, well-known critics such as Barr Ferree did concede that "there has really been developed among us a form of structure which...has...a character of its own sufficiently definite and distinctive to make it an American type. This is the high office-building..."[15] To Ferree the most important aspect of this new building type was its commercial purpose. "The high building is neither a fashion nor a fad; its popularity rests upon the successful manner in which it fulfils an economic necessity in current affairs." Ferree's praise for the skyscraper's utilitarianism formed the foundation on which other writers, both specialist and popular, based their discussion about the aesthetics of the tall building and its contribution to the positive distinctiveness, the American-ness, of New York. The city was the location of the greatest number of such buildings and the place thus most closely associated with the new verticality of the American city.[16]

Discussions of the new skyscrapers revolved around three major themes: their utility, their "American-ness," and their aesthetic appeal. All three themes fit with contemporary American notions of aesthetics in architecture and the decorative arts. Emerging contemporaneously with, but in opposition to, the Beaux Arts style popular at the close of the nineteenth century (the latter exemplified in the 1893 Chicago World's Fair), the American Arts and Crafts movement gradually gained influence during the prewar years, articulating an aesthetic committed to the use of local materials for building, furniture, and housewares and to notions of beauty grounded in a context of functionalism and utility. By contrast, in Europe during this period the early Cubism of Picasso and Braque began to shatter seemingly unquestionable and "natural" forms of representation. Other painters, sculptors, and photographers were developing the works of a European avant-garde modernism that did not come to the attention of the general American public until the Paris Exposition of 1925. Until that time, architects and artists in the United States worked toward an American modernism consciously attempting to construct an aesthetics different from the European past and present.[17]

Among the opinion makers of the weekly and monthly periodical press, a consensus developed in the first decade of the century that the "new New York" exemplified in the new commercial office buildings should continue to chart its own architectural path, and break away from the old Europhilia that devalued American innovations in art and architecture. "[W]hat an accent it gives to the two great highways of the metropolis!" remarked author and critic John Corbin in a 1903 article in *Scribner's Magazine*, referring to the recently opened Flatiron Building, rapidly becoming an icon of New York's modernity and originality.[18] His answer to those critics who objected to the "American" qualities of the building was to indicate the Madison Square Garden building, just across the square from the Flatiron, which he described as "an intelligent variation upon the far-famed campanile of Seville." Corbin was fiercely critical of this type of imitative architecture, and of the vaunted position it held in the eyes of the public and of architectural critics:

> Yet what has a campanile to do with our past, our present, our future?... It is an eternal monument to the fact that those who made it were not able to work out the life of their own time and place into new forms of beauty. Compared with this exquisite exotic from the Old World, no doubt, the rough young stripling of the New is crude and assertive.... Yet the Spanish tower belongs to an alien people and a vanished age. This twentieth century giant, whether ugly or beautiful, stands on the threshold of vigorous new life and of vast architectural possibilities.[19]

Corbin's discussion of New York, like those of many of his contemporary commentators, placed the city within the emergent new geography of space and time whereby Paris and London, the major Old World cities, no longer represented the location of modernity but instead the location of the past. "In all the great cities of the world there are interesting and beautiful things, but they are things of a past," wrote Corbin, "of manifold tradition, or things of a present that is scarcely distinguishable from such a past. The life here is the life of a present that looks out to a future, infinite in the variety of its possibilities."[20]

Part of the valorization of Manhattan's new tall buildings was that they fit with the city's commercial character and with its physical environment, and that they were therefore appropriate to their purpose and to their surroundings. Commenting on the new 625-foot Singer Building, the tallest building in the city and in the nation at the time of its completion in 1908, journalist George E. Walsh both dismissed the skyscraper's one European rival in terms of height, and claimed a "oneupmanship" for the Singer building based on its canny American practicality: "[The Singer Building] is a good deal short of the Eiffel Tower, but that was not constructed as an office-building or living place. It was merely a freak for temporary advertising purposes of a great fair. The tower of the new wonder of the New World will be sixty-five feet square, and on each floor there will be ample office-rooms." Walsh and others celebrated New York's tall buildings for their combination of sheer size and utility, of form and function.[21]

Utility was "the governing consideration" of the size and the design of skyscrapers, commentators argued. Engineering problems had hampered the development of tall buildings but improvements in fireproofing and elevator technology had substantially solved those issues. The solution of such problems had allowed architects more time to consider the buildings' exterior and interior decoration. Such buildings provided modern office space, an effective advertisement for the corporate owner, often a public passageway from one street to another, and shops and services that meant their tenants rarely needed to venture outside the building during the day. The skyscraper, unknown in Europe, was an "American type," wrote engineer Herbert Wade, a "building based on pure utility and special conditions" with "an artistic design and treatment...that to-day justly earns the admiration of European critics".[22]

Architectural critics and commentators almost always linked the utility of the skyscraper to the idea that such buildings were both uniquely American and, increasingly, aesthetically appealing. The pseudonymous "A.C. David" provided perhaps the clearest example of the relation, in the minds of contemporary architectural critics, of commercial utility, Americanism, and aesthetics. David's 1910 article in *Architectural Record* addressed several new commercial buildings on Fourth Avenue between Union Square and 30th Street. The buildings had been constructed to provide office space, warehousing, and showroom space for some large manufacturing and importing companies. David praised the design of the buildings to fit a strictly commercial function. He regarded these examples of commercial architecture as not only distinctly American but as a kind of natural evolution from the local environment.[23] He referred to the buildings as "a normal and natural growth," free from the "perverting and...corrupting" effects of European forms. David closed his article with a Spencerian interpretation of the success of these buildings, claiming that "They are absolutely a case of the survival of the fittest—the fittest, that is, under existing conditions.... [A]ny future advance of American commercial architecture will depend upon a further development of the ideas and the methods which have made these Fourth Avenue buildings what they are." Assessments such as David's built on the association of commerce with both national identity and with New York, thereby implying that these commercial buildings located in Manhattan were both American and natural, an almost organic expression of New York's metropolitan status.

While the pundits and boosters commented on New York's status, more adventurous souls explored and described the nation's less traveled areas "out west." Though located at least two thousand miles west of the Hudson, similar concerns preoccupied these journalists and travelers eager to recommend the West as a tourist destination. Like their Eastern counterparts, the intrepid interpreters of the West during the pre–World War I period went in search of national symbols but in the more familiar arena of such symbols, the land. During the first two decades of the twentieth century, the architectural boosters of New York increasingly linked the "national" landscape of the Far West to the burgeoning "Americanism" of the city's skyscrapers.

Scholarship on the relation between nationalism and landscape imagery places this early twentieth-century American representation of Western landscapes within a long European tradition. Geographer Stephen Daniels has described how through the words and images of poets and painters, often co-opted by the state, particular landscapes become part of nationalist iconographies.[24] In early twentieth-century America, both art and commerce designated the Far Western landscape as the icon of American national identity. In America, as in Europe, particular landscapes have at times better served the purpose of depicting the nation. As Angela Miller has shown, the role played by the mid-nineteenth-century New York school of landscape painters, during a period she refers to as one of "romantic nationalism," was based on the power of the northeast region to commandeer ideas of nationhood, especially during the Civil War period.[25] Much of what Miller describes as constitutive of this period of nationalism and its popular landscape representations bears a striking resemblance to the factors defining the early twentieth century. There is a parallel between the relation of mid-nineteenth-century landscape imagery and national economic prospects as described by Miller, and a similar set of relations at the turn of the twentieth century. Both periods witnessed the nationalist celebration of both the sublime landscape of the wilderness and that of a burgeoning capitalist market. In the early years of the twentieth century, efforts to incorporate the seemingly unnatural and incomprehensible environments of the Far West and of Manhattan into a nationalist iconography brought together these two otherwise dissimilar landscapes. Such efforts formed a necessary stage in the development of a twentieth-century American national identity that needed and used the "landscapes" of both skyscrapers and mountains to further American economic and cultural power at home and abroad.

The concerns of the early twentieth-century chroniclers of the Far West echoed those of the cities trying to describe and categorize the "new New York" during the same years. In search of national symbols, they looked to the deserts and mountains for that which distinguished those lands from Europe. In landscape as in urban aesthetics, Europe remained the yardstick against which all was measured. Like the boosters of New York's emerging skyscraper cityscape, the promoters of the Western landscape had to retrain themselves and their readers/viewers in order to perceive these areas as "beautiful" and worthy of the sometimes arduous trip to reach them. By looking for what was distinctly American in the Western lands they freed themselves from a European landscape aesthetic and thus made possible a new way of seeing the land and incorporating it into the nationalist cultural aesthetic demanded by America's changing international status. Their efforts to nationalize the Western landscape by reevaluating it and redescribing it with a consciously American vocabulary ultimately provided the metaphorical language with which to certify the American-ness of the unnatural landscape of New York City.

Standing in a settler's ramshackle hut, "face to face with the bare desert," Harriet Monroe, a genteel traveler in Arizona at the turn of the century, cast her mind back to her travels of the previous two summers. Monroe felt "ashamed" of her former European preoccupations as she struggled with her response to the Arizona desert: "It was a most complex emotion, this vision of unachieved glory set against a background of immemorial antiquity. For the desert is old beyond one's dreams of age; it makes Rome or Nineveh seem a thing of yesterday."[26] She thus described her initially disquieting but important realization that America possessed a history of its own, legible in the lands of the Southwest, still largely unsettled by towns or cities whose newness might detract from the dignified "antiquity" of the desert. Monroe's observation, however, also proposed the role of that region and landscape in an even more glorious future.

Monroe's response to the American Southwest characterizes that of many other travelers and journalists exploring this region in the early years of the twentieth century. Accustomed to equating the aesthetics of landscape with the countryside and towns of Western Europe, early twentieth-century travelers in the American West and Southwest struggled to reconcile a previous generation's ideas about the ugliness and barrenness of those regions with their own perceptions of the apparent beauties of these lands. What had changed to enable Americans to perceive the desert and canyon lands, once deemed barren and irredeemably ugly, as at least as beautiful as the landscapes of Europe?

America's defeat of Spain in the war of 1898 marked the United States' entry into the elite club of imperialist nations. The new sense of national unity provoked by the war came shortly after the very practical uniting of the nation's coasts and interiors by the railroad magnates. The development of railroad connections through the Southwest into Southern California and up the Pacific Coast provided the last links in the iron roads linking industry and settlement throughout the West. The railroads provided both the access and many of the amenities necessary not only for industry but also for tourism in these previously little traveled regions. The period of the greatest expansion and redevelopment of the nation's railroads, 1890–1917, coincided with, and no doubt contributed to, the growing popularity of the West as a tourist destination.[27] The transcontinental railroad companies all advertised excursion trips to the West, and adventurous Easterners could buy a complete package of train ticket, hotel accommodation, and sightseeing tours, all provided by the railroad companies. Railroad companies constructed some of the West's finest hotels in what became and remained the most popular tourist destinations in the country—such as the Grand Canyon, Yellowstone, and Yosemite National Parks.[28] A new generation of explorers of the Western states and territories, borne in relative comfort to their destinations, made a vital contribution to this early phase of American cultural nationalism by imitating a well-established European method of boosting national identity and national pride: marshalling the nation's landscape as the embodiment of national values, strengths, and characteristics.

Travel writers and tour promoters increasingly represented the landscapes of the American West and Southwest in the first years of the century as the great American landscape, the ultimate natural representation of American identity. Echoing the contemporaneous claims made for the metropolitan and aesthetic qualities of New York City, boosters claimed that the landscapes of these regions differed from—but aesthetically equalled or bettered—those of Europe. Harriet Monroe, in her account of her impressions of Arizona referred to above, reprimanded herself and other Americans for preferring "months or years of wandering in history-haunted Europe" rather than exploring the apparently distinctly American landscapes of the Southwest. The challenge, she and other writers suggested, was not only to perceive the landscape's specifically American beauty, but to wean the touring classes away from their European trips and bring them West.[29]

The sublime beauties of the undeveloped American West differed from the pastoral attractions of Europe's settled landscapes. Most middle- and upper-class Americans had more familiarity, either from personal travel experience or from landscape painting, with European standards of natural beauty. Robert Hill, of the United States Geological Survey, describing the "American Desert" to readers of *World's Work* in 1902, referred to the region as "a strange land of paradox, where each rock and tree and flower and river reverses conventional tenets and laws and conditions for Anglo-Saxon environment, as founded upon ideas preconceived by thousands of years of ancestral experience." He described the "aberrant" features of the desert regions that included many areas "which are apparently a mockery of nature."[30] Hill responded to an apparently "unnatural nature" whose aesthetic appeal he could neither deny nor yet describe.

Although familiar, the beauties of the European pastoral landscape had become, Monroe and others suggested, too comfortable, offering no excitement or challenge. Of Italy Harriet Monroe wrote that "[i]ts beauty is self-contained and measurable; one rests in it with profound content," whereas in Arizona "Nature is not conciliatory and charming; she is terrible and magnificent." Such sublime landscapes offered an "unfamiliar and incomprehensible beauty" to Americans raised on Western European standards of pastoral beauty.[31] "One accustomed to the lovely scenery of England, to the soft, caressing beauty of her landscape, is ill prepared to grasp the vastness of the Yosemite, the Yellowstone or the Grand Canyon of Arizona," wrote Robert Hill. "We love what we can understand," Monroe continued, "what history and letters and art have taught us to understand . . . We prefer to follow other feet,—to see Shakespeare's England and Byron's Switzerland; But no poet has said an adequate word for these unexplored sublimities; . . . to them the mind of man must venture as a pioneer; . . . "[32] These chroniclers of the West had discovered an American sublime landscape, whose sublime status resided, paradoxically, in its seeming unnaturalness. Their responses echoed those of early twentieth-century observers of Manhattan's massed skyscrapers. Both landscapes seemed initially to lack the familiar components of beauty, yet commerce and art ultimately reconciled both with a new, distinctly American aesthetic in the service of national identity.

As Anne Farrar Hyde has argued in her study of the relation between the Far Western landscape and American national culture at the turn of the century, as American travelers became more familiar with, and enamored of, the appearance of the arid plains, canyon lands, and deserts of the American West, they acquired a language with which to describe it.[33] Hyde argues that early twentieth-century American writers and artists moved away from a descriptive language reliant upon European comparisons and aesthetic standards. Writers less frequently described the mountain ranges, desert rock formations, and climates of the West as American versions of Swiss Alps, or as akin to the turrets and spires of European castles and cathedrals, or as an American Italy.[34] The West, according to Hyde, became "defiantly American," as writers struggled to describe the mountainous and arid lands in a language more appropriate to its newfound national (and potential commercial) value.[35]

However, seeking symbols of national identity in the natural landscape was not an especially "American" impulse. Natural landscapes and geological formations were inextricably tied to constructions of national identity and culture in Europe. England's "white cliffs of Dover," Germany's Bavarian forests, and Switzerland's Alps had all been marshalled by European nations and governments to help define a sense of nationhood and national identity linked to nature. Such images helped to "naturalize" the often political concepts of nation and nationality. Moreover, the language used to describe the American natural landscape in a supposedly more "American" way was, of course, Spanish–the terms "sierra," "cañon," "mesa," and "cordillera" had been applied to the region by the previous (European) colonizers of the West and Southwest.[36]

Nonetheless, the seemingly undeniable American-ness of the Western landscape due to its apparent reversal of European natural landscape forms, and its soaring popularity with Eastern tourists, made the language of this landscape the ideal to which to compare the newfound American-ness of the skyscraper city. If the mesas, sierras, and canyons of the Western states represented what distinguished America from Europe, and what gave Americans an indigenous landscape and an ancient indigenous past, then what better way to describe and represent the American-ness—the national cultural value—of the skyscraper city of New York but in the language of the natural landscape of the Far West?

Upon arriving in New York City in 1912, French author Pierre Loti felt disturbed at how different the city seemed from Paris.

Observing the city from his hotel room, he remarked: "When I return from here, Paris will seem just a quiet, old-fashioned little town."[37] To the Frenchman, New York seemed an "infernal abyss," "almost a nightmare."[38] Loti, accustomed to the ordered uniformity of Haussmann's Paris, expressed consternation at the sight of the city's skyscrapers: "[t]hey rise up here and there, as though by chance, alternating with normal... buildings;... they seem like houses that have caught a strange disease of over-growth and madly shot up to distorted heights."[39] A few years later, however, an anonymous visitor from Laramie, Wyoming had a very different view of the city. Describing his first impressions of New York, this Westerner remarked, "When I came to New York... I liked the skyscrapers... They helped relieve my loneliness at the absence of mountains."[40] Both visitors responded to the massed height of Manhattan's skyscrapers and their haphazard arrangement in the lower part of the city, one with horror, the other delighting in the appearance of familiarity.

How could the two visitors have had such different perceptions of New York's built environment? The Westerner saw "mountains" where the Parisian saw "modernism." The Westerner's claims fit with a new urban discourse that developed during the first decade of the twentieth century that placed the city—in particular New York City—at the heart of a visual and textual rhetoric about the meaning of America, a discourse that sought national symbols for a contested national identity. The visitor from Wyoming offered his readers what, by 1917, had become a familiar trope: Manhattan as a version of the Far West's rocky landscape. Naturalizing the skyscrapers—icons of New York's and America's modernity—Americanized New York. The city was thus made an acceptable national metropolis for a people hungry for nationalist symbols, but not yet ready to see themselves represented by New York's urban skyline.

Loti's statement suggests that New York's status as "the capital of modernism" resituated Paris and New York in both spatial and temporal relationship. By thus describing New York, Loti designated the city as the locus of the present, of current trends. In comparison to New York, he says, Paris will now seem "old-fashioned," the location of the *passé*. Geographer James Duncan has described this process as one whereby a "journey in space is a journey in time." Addressing what he calls "sites of representation," Duncan shows how geographical areas are understood to bear temporal properties of past, present, or future. In early twentieth-century America, the representation of Paris as the "past" and New York as the "present" or "future," which repositioned

New York (America) in relation to Paris (Europe), was achieved by means of the Americanization of both New York's built environment and Western America's natural landscape.[41]

Both skyscrapers and mountains proved necessary to the development of an American New York. Civic boosters in the city government and the business community interested in promoting New York's claims to metropolitan status needed first to establish a larger American national identity and set of symbols upon which to base those claims. This work necessitated a cultural separation from Europe through the discovery and celebration of a unique American identity appropriate to an industrialized, urbanizing nation. Early twentieth-century American journalists, photographers, tour promoters, and government officials identified an indigenous, authentic American identity in the natural landscapes of the American West, in the land formations, geology, and climates that most differed from those of Western Europe. This generation of travelers similarly attributed to the native peoples of the West and Southwest a natural, indigenous, and authentically "American" beauty. Like the landscapes they inhabited, white observers had previously regarded native Americans as ugly and not worth looking at.

The "discovery" and representation of the West as "America" provided a national imagery, language, and historical foundation on which to base the construction of an acceptably distinct American present. Until the middle of the twentieth century, the ultimate manifestation and location of that American present was New York City. It was therefore vital to Americanize New York by interpreting Manhattan, in word and image, as a natural "American" landscape.[42] Representations of American Western landscapes were not entirely new to Eastern Americans in the first decades of the twentieth century. Photographs by Timothy O'Sullivan, Carleton Watkins, John Hillers, and William Bell had become familiar from the widely popular, published government surveys of the West in the late nineteenth century.[43] These images, as well as the paintings of Albert Bierstadt, Thomas Moran, and Frederick Remington, had depicted the Rocky Mountains, deserts, salt lakes, hot springs, and geysers of the Western states and territories. What changed in the twentieth century was, first, the incorporation of these landscapes into a confident cultural nationalism and, second, the ability of a greater number of Americans to visit these areas for themselves as tourists.

At the close of the nineteenth century, the language of the Western landscape had been applied pejoratively to the cityscapes of New York and Chicago to describe narrow streets dominated by new tall buildings, and to imply the ill effects of the new commercial capitalism on older ways of urban life. Henry Blake Fuller famously described the new middle-class apartment dwellers and office workers of 1890s Chicago as "cliff dwellers" in his novel of that name.[44] He opened his story of the harshness and immorality of urban capitalism by describing Chicago's changing urban landscape as a "tumultuous territory through which . . . the rushing streams of commerce have worn many a deep and rugged chasm."[45] The early years of the twentieth century saw the use of that language shift to construct a celebration of the aesthetic appeal, and the national and international significance, of New York City.

The burgeoning tourist industry provided the most important connection associating images of the West with images of New York. Guidebooks, maps, stereographs, postcards, and bus tours of cities and natural "wonders" constructed a coherent set of images of the nation, ordered in a sequence that allowed the visitor or viewer to see the major designated "sights" of each place within a reasonable amount of time—whether 30 minutes to view a set of stereographs or a week spent visiting a particular place of interest. The appeal to tourists, domestic and foreign, of cities such as New York or areas such as the Western states was the "proof" of the promotion of such sites as representative of the nation.

Guidebooks, postcards, and stereographs proudly associated New York's new urban forms with Western landscape features. "New York Skyscrapers," the first section of a 1902 New York guide, described the unique impression made by the high buildings as one approached the city from the bay. The impact of "these architectural marvels" was further enhanced, the author stated, "as we wander through the downtown streets, and passing from one shadowy cañon into another make our way between the tremendous cliffs."[46] A 1905 pamphlet of New York views distributed as advertising material by the Singer Sewing Machine Company included a photographic view of Broadway, looking south, taken from the St. Paul Building, named the "Grand Canyon."[47] The photograph, by George P. Hall and Company, was reproduced many times in other guidebooks, in stereograph series, and in postcards. Images of downtown Broadway and other streets in Lower Manhattan were routinely captioned as "canyons" in booklets of views issued by guidebook companies or hotels.

The visual trend for photographing the lower part of the city from the roof of a skyscraper or in horizontal format contributed to the perception of New York as a landscape. A reading of the stereograph catalogues issued by two of the largest distributors of such views, Underwood and Underwood and H.C. White and Company, in the early years of the century, demonstrates that the vast majority of the stereograph views of New York City consisted of images taken from atop the World Building, the St. Paul Building, the Flatiron, the Singer Building, the Metropolitan Life tower and, a little later, the Woolworth Building. From these vantage points, the city appeared in almost panoramic perspective as a landscape of clusters of skyscrapers set among older, seemingly squat, stores and warehouses. Part of the appeal of these rooftop views, like those recommended in tourist guidebooks, was the suggestion that one could now see this fantastic cityscape from a perspective previously only imagined. The perspective thus accentuated the suggestions that New York, like the canyon lands and mountains of the West, constituted a sublime landscape. The streets bordering the buildings seen from such a sharp angle might indeed have seemed to resemble the narrow, deep canyons familiar from descriptions and images of Colorado and Arizona.

As William Taylor has argued, the horizontal skyline view of New York marked an iconographic watershed in the development of ways of seeing the city.[48] Surrounded by water and at the entrance to one of the world's great oceans, New York had always lent itself to such horizontal views, often from across one of the rivers or from the bay. The development of the new tall buildings compounded the preference for this view of the city. The skyline view offered an image of New York that clearly associated it with the visual conventions of Western landscape painting and photography. Panoramic images taken from photographs became regular features of guidebooks and photographic souvenirs of the city. Taken from a vantage point either in Brooklyn looking across the East River or from across the Hudson, these images presented a jagged horizon of tall buildings, usually identified by name at the base of the image. Water below and sky above frame the long narrow strip of land dominated by a series of variously shaped "summits." The skyline image of the downtown skyscrapers removed them from the larger context of the still predominantly low-storied city, creating the impression that this architectural landscape typified the city.

Both these early twentieth-century images, the rooftop view and the skyline, differed from the older "bird's eye view" images of New York popular in the late nineteenth century by their angle of view, which was much lower than the older images. The bird's eye views had depicted the whole city, from an almost vertical imagined position in the sky above the Bay. While they included the important buildings and institutions of the city, often exaggerated in proportion to other buildings, such features of the bird's eye views were merely part of an overall view of the city. The rooftop panoramas situated the viewer in the midst of the major business section of the city, erasing any larger geographical context. The skyline image of the early twentieth century similarly emphasized the city's commercial architecture, making clear its dominance over the rest of the city.

Both the nineteenth-century bird's eye views and the early twentieth-century skyline images belonged to the genre of "booster" images designed to promote a particular place, for the purposes of settlement, real estate, or tourism. As such, these images did not include visual clues to other aspects of the place depicted that might detract from this idealized impression. For example, like the older bird's eye views, these new images made invisible the varied populations occupying both the buildings and the hectic streets that surrounded them. The skyscrapers in these early twentieth-century images appeared to form a monumental, unpopulated, silent region, similar to the romanticized landscape paintings of the American West by such artists as Albert Bierstadt and Thomas Cole. For tourists to New York City from the West, and for Easterners familiar with images of the Western landscape in their monthly magazines, the skyline image of New York fit into a developing visual lexicon of the American landscape in the early twentieth century—a land of cliff dwellings, mesas, canyons, and sierras.

The jagged irregularity of New York's skyline, a product of the city's lack of zoning and height restrictions in the pre–World War I period, contributed to the seeming "naturalness" of the skyscrapers and the possibility of seeing them collectively as a landscape.[49] Financial considerations, such as proximity to other relevant businesses and financial institutions, access to transportation, the availability and cost of a building plot, as well as the whims and wealth of the building's owner determined the location, height, and style of these office buildings. Tall buildings in lower Manhattan emerged, therefore, without regard to the overall appearance or organization of this section of the city or of Manhattan as a whole. This very visible part of the city thereby acquired a haphazard look that some observers used to reinforce their representations of the city as both beautiful and as "natural," as a landscape.

While some complained that New York would never be as beautiful as European cities because of its commercial character and lack of planning, others saw these very features as part of New York's nascent "natural" beauty. They argued that the skyscraper had developed and grown in New York to meet local needs; it was, therefore, "natural," and should not be bound by the "unnaturalness" of strict urban planning. In his 1903 *Scribner's* article, illustrated with photographs by Alfred Stieglitz, John Corbin argued that New York's irregular appearance was actually part of its particular beauty. What the city needed, he argued, was not a restructuring or redesign, but rather a new way of looking at and representing this urban beauty.[50]

Corbin saw beauty in the city but struggled to find a language and set of metaphors to express what he saw. Stieglitz's pictorialist photographs exhibit a similar tension between an appreciation of the city's aesthetic potential and an unwillingness to depart from older stylistic conventions. Corbin appreciated the novel visual effects of the new architecture on the city's form and appearance—the "surprises of perspective" and the "juxtaposition of masses"—yet employed both European and American landscape metaphors to describe the aesthetic appeal of Lower Manhattan seen from the Bay. He described the streets as "canyons," the skyscrapers as "cliff-like," and their collective appearance at sunset as resembling the beauty of "a snow-capped Alp." Corbin's metaphoric confusion, similar to that expressed by his contemporaries observing the Far West, suggested that New York's urban landscape also existed in the midst of an enthusiastic but as yet incomplete Americanization.[51]

Four years later, with the city's skyline joined by the Times Building in Longacre (thereafter renamed Times) Square, and the Singer Building rising up on Broadway and Liberty Streets downtown, the idea grew amongst the city's boosters that New York's skyscrapers gave the city a "natural" beauty. Others echoed Corbin's suggestion—represented in Stieglitz's photographs—that the city's beauty appeared most evident seen through a soft-focus lens, a mist, in some sort of half-light, or in a snow storm. Such effects of light and weather softened the outlines of the buildings, and merged the assorted colors of the city into one or two tones, emphasizing overall shapes and forms, rather than the sometimes jarring details of the city and its inhabitants. Such a pictorialist view of the city made it appear more "natural," more part of a landscape.

A 1907 article in *The Craftsman*, the mouthpiece of the American Arts and Crafts movement, exemplified this discursive trend, celebrating the haphazard appearance of the city's skyline. Addressing a recently published set of etchings of New York skyscrapers by the well-known artist and illustrator Joseph Pennell, the article was entitled,

"How New York Has Redeemed Herself from Ugliness—An Artist's Revelation of the Beauty of the Skyscraper."[52] The author, Mary Fanton Roberts, a frequent contributor to the magazine under the pseudonym "Giles Edgerton," opened her article by describing an approach to New York from across the Bay. Edgerton referred to the "uneven lines" of the city's horizon, and to the "canyons" stretching between them. She made the important point that when an observer viewed the individually dissimilar and unrelated buildings in aggregate they acquired a "picturesque" appearance. The architectural picturesque, she argued, could only come about "in the wake of a real need expressed with intelligence and skill."[53] Like many of her contemporaries, Edgerton connected the utilitarianism of the New York skyscrapers (in terms of their function and their appropriateness to a city with limited land on which to build) to the notion that they were natural to the local environment, and therefore American. From this perspective it seemed a logical step to see the city as an American natural landscape, comparable in form and cultural value to the Far West.[54]

The image of New York as an all-American landscape, an architectural version of the rocky West, erased from view the day-to-day tensions of the city at street level. In particular, this Western, Americanized version of New York made invisible the city's growing immigrant population. During the time the city's boosters and other image makers had promoted this architectural reading of the city, approximately fifteen million immigrants, mostly from Southern and Eastern Europe had arrived in New York. Although not visible in the skyline of the city, or even from many of Manhattan's skyscrapers' windows, these newcomers profoundly reshaped the city's popular culture, politics, and ethnic makeup by the 1920s. Furthermore, the city's immigrant presence put some members of New York's Anglo elite on the defensive, feeding the flames of anti-Semitism and popularizing scientific racism. These anti-immigration elites and their middle-class supporters around the country worked to re-spin older ideas and images of New York as un-American. This time, the charges against New York's Americanism were based less on nineteenth-century antiurbanism and more on renewed racism and conservatism in the post–World War I period. With the most striking phase of architectural innovation completed by 1920, and public concerns over American-ness focused on bodies more than buildings, the city's built environment played a subordinate role to its population in determining New York's national meaning during the 1920s.

Notes

This chapter is adapted from a chapter in my book *How New York Became American, 1890–1924* (Baltimore: Johns Hopkins University Press, 2006).

1. The clearest representation of this phrase and its meaning can be found in John C. Van Dyke, *The New New York* (New York: Macmillan, 1909).
2. Max Page has referred to this process, with reference to Baron Haussmann's rebuilding of Paris, as a process of "creative destruction." See Page, *The Creative Destruction of Manhattan, 1900–1940* (Chicago: University of Chicago Press, 1999).
3. Jules Vallée Guérin (1866–1946) was born in St. Louis, Missouri, and began exhibiting his work at the Art Institute of Chicago in the 1890s. He moved to New York in 1900 and there made his name as the leading painter of architectural subjects and as an illustrator. His lithographs appeared in numerous articles in *Century, Scribner's*, and *Harper's*. As an architectural renderer he worked with leading architects of his day on major projects, including Charles McKim's winning plans for Washington, DC's Mall, and with Daniel Burnham and Edward Bennett on their "Plan of Chicago," 1909.
4. Randall Blackshaw, "The New New York," *Century Magazine* 64 (August 1902): 492–513, 493.
5. The literature on Paris as the capital of France is enormous; that on London less so. For Paris, see David Pinkney, *Napoleon III and the Rebuilding of Paris* (Princeton: Princeton University Press, 1958); David Jordan, *Transforming Paris* (New York: Free Press, 1995); David Harvey, *Consciousness and the Urban Experience* (Baltimore: Johns Hopkins University Press, 1985). For London, see Donald Olsen, *Growth of Victorian London* (New York: Holmes and Meier, 1976); Ken Young and Patricia Garside, *Metropolitan London: Politics and Urban Change, 1837–1981* (New York: Holmes and Meier, 1982).
6. Herbert Croly, "New York as the American Metropolis," *Architectural Record* 13 (March 1903): 199–200, 205. See other articles by Croly, such as "American Architecture of To-Day," *Architectural Record* 14 (December 1903): 413–435; "New World and the New Art," *Architectural Record* 12 (June 1902): 134–153; "What Is Indigenous Architecture?" *Architectural Record* 21 (June 1907): 434–442.
7. Hamilton Wright Mabie, "The Genius of the Cosmopolitan City," *The Outlook* 76 (March 5, 1904): 577–593, 578.
8. Ibid., 588, 593. As historian John Higham has noted, 1906–1907 marked "a new phase in the history of American nativism." Higham, *Strangers in the Land: Patterns in American Nativism, 1860–1925* (New York: Atheneum, 1977), 165. See also Stephen J. Diner, *A Very Different Age: Americans of the Progressive Era* (New York: Hill and Wang, 1998), 77.
9. The attraction of New York for the newly wealthy was not always popular with longer established New Yorkers, despite the wealth such individuals or their corporations brought to the city. Like the character Dryfoos in William Dean Howells' novel about class and money in late nineteenth-century New York, *A Hazard of New Fortunes*, the new millionaires from the Midwest and West were frequently depicted as *arrivistes*, as flashy buffoons. Their new residences on Fifth Avenue were disparaged by the architectural press as "ridiculous" for their ostentation and poor taste. See Franz K. Winkler, "Architecture in the Billionaire District of New York City," *Architectural Record* 11 (October 1901): 679–699. This scathing article contains photographs of, and commentary on, the new homes of such men as Andrew Carnegie and F.W. Woolworth. Howells' novel, although published in 1890, still provided a relevant description of the tensions between old money and new money families as they negotiated their respective places in the city's increasingly complicated social hierarchy.
10. Ibid., 590.

11. For a discussion of the growth of white-collar work in New York's financial sector during these years, see Angel Kwollek-Folland, *Engendering Business: Men and Women in the Corporate Office* (Baltimore: Johns Hopkins University Press, 1994), 15–40.
12. Carol Willis argues for an economic interpretation of New York's skyscrapers in her *Form Follows Finance: Skyscrapers and Skylines in New York and Chicago* (New York: Princeton Architectural Press, 1995). Cultural geographer Mona Domosh offers a spatial and cultural argument for the building and significance of New York's skyscrapers in her *Invented Cities: The Creation of Landscape in Nineteenth Century New York and Boston* (New Haven: Yale University Press, 1996).
13. See William Taylor, *In Pursuit of Gotham: Culture and Commerce in New York* (New York: Oxford University Press, 1992), 23–33 and 51–67.
14. Joseph B. Gilder, "The City of Dreadful Height," *Putnam's Monthly* 5 (November 1908): 131–143, 132.
15. Barr Ferree, "The High Building and Its Art," *Scribner's Monthly* 15 (March 1894): 297–318, 297.
16. Ibid. In the introduction to this project I discuss the contrasting national role and image of Chicago, the other American city associated with early skyscraper architecture.
17. English aesthetic theorists John Ruskin and William Morris initiated what became the Arts and Crafts movement in England. The ideas underpinned the American version of that movement in the early twentieth century. The leading figure in the American Arts and Crafts movement was Gustav Stickley. On the philosophy and cultural politics of the American movement, see Eileen Boris, *Art and Labor: Ruskin, Morris, and the Craftsman Ideal in America* (Philadelphia: Temple University Press, 1986); Robert Judson Clark, ed., *The Arts and Crafts Movement in America, 1875–1920* (Boston: Museum of Fine Arts, 1987).
18. John Corbin, "The Twentieth Century City," *Scribner's Magazine* 33 (March 1903): 259–272, 261.
19. Ibid., 262. Not all architectural critics agreed with the idea that American architects should break away from the building styles of Europe. The architect A.J. Bloor seemed to address Corbin directly when he wrote, "Who, of any taste in architecture, regrets that a beautiful tower in Seville should be virtually reproduced in Madison Square Garden?" See Bloor, "A Letter on Current American Architecture (Including the 'Skyscraper') and Architects" (Pamphlet, n.d., c.1905). Footnote to title reads: "Suggested by an article of Mr. Montgomery Schuyler, in the 'Metropolitan Magazine' of July 1905." Warshaw Collection, Archives Center, NMAH. Box: "Architecture.") An editorial in *Scientific American* in March 1907, entitled "A City of Towers," criticized the increasing numbers of tall buildings in New York. If, however, there were to be tall buildings in the city, the editors of the periodical suggested that architects work to accentuate the vertical rather than the horizontal lines of the buildings. They suggested taking "one of the beautiful cathedral towers of Europe as a model," to "reproduce something of the effect the great Gothic windows and other characteristic effects of these handsome structures." See "City of Towers," *Scientific American* 96 (March 30, 1907): 266. Six years later, Cass Gilbert, architect of the Woolworth Building, applied just such a Gothic treatment to the corporate offices of the "five and dime" millionaire.
20. Ibid., 264, 266. The only European city to which Corbin ascribed a current modernity was Vienna, and he did so while arguing that the women of New York were more fashionable and more stylish than the women of Paris and London. Corbin noted that to find her European counterpart one would "have to go to the Kohlmarkt and the Ringstrasse. In Vienna the life is similarly shifting and cosmopolitan, there is a similar lack of indigenous style, and a similar willingness to take the best, wherever it is to be found."
21. George Ethelbert Walsh, "Modern Towers of Babel in New York," *Harper's Weekly* 151 (January 12, 1907): 68. See other articles by Walsh, such as, "Potential Value of a City

Roof," *World To-Day* 10 (June 1906): 646–647; "Traffic Congestion in New York," *Cassier* 34 (June 1908): 151–155. The planning, building, and completion of the Singer Building, the world's tallest building until the completion of the Metropolitan Life Tower in 1909, occasioned many articles on the merits and styles of tall buildings in New York. See, for example, "Office-Building 612 Feet Tall," *Scientific American* 95 (September 8, 1906): 169, 174; C.M. Ripley, "Singer Building, New York, Forty-Seven Stories High," *World's Work* 14 (October 1907): 9459–9461; "Erection of the 612-Foot Singer Building," *Scientific American* 97 (September 7, 1907): 168–169; Gardner Richardson, "Great Towers of New York," *The Independent* 65 (August 6, 1908): 301–305.

22. Herbert T. Wade, "Tall Buildings and Their Problems," *American Review of Reviews* 38 (November 1908): 577, 586. See also F.W. Fitzpatrick, "Building Against Fire," *Outlook* 88 (April 25, 1908): 936–945; J.K. Freitag, "Fire Prevention in High Buildings," *Engineering Magazine* 34 (February 1908): 735–740; T.K. Thomson, "Caisson Foundations of Skyscrapers," *Scientific American* 65 (March 7, 1908): 152–154.
23. A.C. David, "The New Architecture: The First American Type of Real Value," *Architectural Record* 28 (December 1910): 388–403.
24. Stephen Daniels, *Fields of Vision: Landscape Imagery and National Identity in England and the United States* (Princeton: Princeton University Press, 1993), 5.
25. Angela Miller, *Empire of the Eye: Landscape Representation and American Cultural Politics, 1825–1875* (Ithaca: Cornell University Press, 1993), 7.
26. Harriet Monroe, "Arizona," *Atlantic Monthly* 89 (June 1902): 780–781.
27. For histories of the railroad system and its effects on national culture, see Sarah H. Gordon, *Passage to Union: How the Railroads Transformed American Life, 1829–1929* (Chicago: Ivan R. Dee, 1996); John F. Stover, *American Railroads* (Chicago: University of Chicago Press, 1976). For the West specifically, see the entry under "railroads" in Howard R. Lamar, ed., *The Reader's Encyclopedia of the American West* (New York: Thomas Y. Crowell, 1977).
28. For discussions of the role of the railroads in building Western tourism and developing National Parks, see Ann Farrar Hyde, *An American Vision: Far Western Landscape and National Culture, 1890–1920* (New York: New York University Press, 1990); Marguerite Shaffer, *See America First*, especially Chapter 6; and Alfred Runte, *Trains of Discovery: Western Railroads and the National Parks* (Niwot, CO: Robert Rinehart, 1990).
29. Monroe, "Arizona," 780.
30. Robert T. Hill, "The Wonders of the American Desert," *World's Work* 3 (March 1902): 1821–23.
31. Monroe, "Arizona," 781.
32. Ibid., 782.
33. Hyde, *American Vision*.
34. For an example of the metaphorical muddle brought on by efforts to describe the unfamiliar landscapes of the West, see Arthur Inkersly, "The Grand Canyon of Arizona," *Overland Monthly* 41 (June 1903): 423–432.
35. Hyde, *American Vision*, 208–209.
36. On the construction of national identities see Benedict Anderson, *Imagined Communities* (New York: Verso, 1983); John Bodnar, ed., *Bonds of Affection* (Princeton: Princeton University Press, 1996); Eric Hobsbawm and Terence Ranger, eds., *Invention of Tradition* (New York: Cambridge University Press, 1983).
37. Pierre Loti, "Impressions of New York," *Century Magazine* 85 (February 1913): 611. Pierre Loti, pseudonym of Louis-Marie-Julien Viaud (1850–1923), a French naval officer, writer, photographer, and artist, became best known in the United States for his play "Madame Chrysanthemum," the basis for the better-known play "Madame Butterfly." He adopted the name Pierre Loti in 1881, taking the last name from the followers of a Tahitian Monarch, Queen Pomaré IV. Loti's Orientalism, stemming from his travels in the South Pacific and the Far East, no doubt influenced his reaction to New York City.

38. Ibid., 609.
39. Idem, "Impressions of New York" (second article in two-part series), *Century Magazine* 85 (March 1913): 758–759.
40. Anonymous, "Confessions of a Westerner and What He Sees in New York," *The Independent* 91 (September 15, 1917): 424, 431.
41. James Duncan, "Sites of Representation: Place, Time and the Discourse of the Other," in James Duncan and David Ley, eds., *Place, Culture, Representation* (New York: Routledge, 1993). For an eloquent and important discussion of the issue of the construction of time, space, and place, see Johannes Fabian, *Time and the Other: How Anthropology Makes Its Object* (New York: Columbia University Press, 1983).
42. See William H. Truettner, ed., *The West as America: Reinterpreting Images of the Frontier, 1820–1920* (Washington, DC: National Museum of American Art/Smithsonian Institution Press, 1991); Beaumont and Nancy Newhall, *T.H. O'Sullivan: Photographer* (Rochester: George Eastman House/Amon Carter Museum of Western Art, 1966); David Margolis, *To Delight the Eye: The Original Photographic Book Illustrations of the American West* (Dallas: DeGolyer Library, 1994).
43. The most recent critical analysis of the role photography played in constructing the imagery and mythology of the American West is Martha Sandweiss, *Print the Legend: Photography and the American West* (New Haven: Yale University Press, 2002).
44. Henry Blake Fuller, *The Cliff-Dwellers: A Novel* (Ridgewood, NJ: Gregg Press, 1968). Originally published by Harper Brothers, 1893.
45. Ibid., 1–3.
46. Foster and Reynolds Company, *New York: The Metropolis of the Western World* (New York: Foster and Reynolds, 1902), 9.
47. Singer Manufacturing Company, *Singer Souvenirs of New York City* (New York: Singer Manufacturing Company, 1905). NMAH. Archives Center. Warshaw Collection. New York City Boxes, Box 21.
48. Taylor, *In Pursuit of Gotham*, 23.
49. Zoning, height restrictions and the requiring of "setbacks" to stagger the mass of tall buildings was not introduced until 1916 and then, more comprehensively, in the early 1920s following the Regional Plan for Greater New York. For more on the history of planning and building laws in New York City, see Richard Plunz, *History of Housing in New York City* (New York: Columbia University Press, 1990) and essays by Keith Revell, Marc Weiss, and Robert Fishman in David Ward and Olivier Zunz, eds., *The Landscape of Modernity: Essays on New York City* (New York: Russell Sage, 1992).
50. Corbin, "Twentieth Century City."
51. Ibid., 260.
52. The publication reviewed was Frederick Keppel, *Mr. Pennell's Etchings of New York "Sky Scrapers"* (New York: Frederick Keppel, 1905).
53. Giles Edgerton, "How New York Has Redeemed Herself from Ugliness—An Artist's Revelation of the Beauty of the Skyscraper," *Craftsman* 11 (January 1907): 458.
54. See also *Architectural Record* 22 (September 1907):161; and Joseph B. Guilder, "The City of Dreadful Height," *Putnam's Monthly* 5 (November 1908): 131–143, pages 131 and 136 especially, for references to the seeming naturalness of the skyscraper landscape, and the appearance from the bay of New York as a city "set on a hill," echoing much older constructions of the meaning and role of America.

PART 3

Antiurbanism in Society and Politics

CHAPTER SEVEN

Imagining the Urban Poor: Poverty and the Fear of Cities

Roger A. Salerno

Before cities there were no poor. In the nomadic world hand-to-mouth existence characterized everyday survival. The less well off lived among the better off. It was only with the emergence of large permanent settlements and cities that a surplus created an institutionalized system of stratification and a category of people referred to as *the poor.* This chapter examines images of the poor: how poverty is associated with urban life and how it is often constructed to stigmatize cities. A brief survey of imagery of the poor in medieval and industrial cities is followed by an examination of more contemporary imagery of the urban lower classes.

Inventing the Holy Poor in Late Antiquity

Michel Mollant tells us that in late antiquity the city of Constantinople was teeming with poor people as opposed to the countryside where paupers were more common, less regulated, and therefore less exposed to public scrutiny. The distinction between the pauper and the poor was that the pauper was often free to own some property. The word most closely associated with pauper was the term *humble.*[1] Most frequently the term pauper was used to describe peasants who tilled the soil and were dominated by estate owners. There on pastoral lands paupers eked out an existence and when their survival was threatened they sought out local sources of help including neighbors and the owner of the estate on which they worked. But if severe economic disasters struck most left the land and headed to the city—there they became poor. In majors cities one could often find formal and informal charity, which helped define who the poor were. As far back as 400 A.D. cities such as Antioch, Constantinople, Alexandria, and Jerusalem recorded very large concentrations of the poor and noted public concerns about them. Reactions ranged from compassion to hostility.[2] Zoticos, a high-ranking official in Constantine's administration, used public money to create a leprosarium in the Galatan hills just outside the city walls. When Constans discovered what he had done for the lepers he had Zoticos drawn and quartered by mules and condemned the lepers to expulsion from their homes and in some cases to death.[3] By the sixth century, a "tidal wave" of poverty swept what remained of the eastern portion of the Roman Empire. Bad harvests, unscrupulous usurers, and mortgage failures were responsible for much of this. Throughout this time the hungry posed a threat to the very small number that maintained their positions of privilege.

Interestingly it was Eastern Christendom—the Byzantine Empire that preceded the West in high concentrations of the urban indigent and therefore methods of their identification and regulation. The Justinian Code had a profound impact on how the poor were envisioned, articulated, and treated. It was to influence European administrations in establishing categories and typologies of indigence: the orphaned, the widowed, the disabled, the diseased, the homeless, the "able-bodied" but unemployed. Here was the first articulation of the "deserving" and "undeserving" poor; here also was a "poverty-line"—an arbitrary economic floor beneath which one needed charity.[4] Each category of poverty was assigned to a particular entity that would look after its interests. And nearly all services to help those in need were administered by the Church. Carolingian monarchs proscribed use of Church property to serve the needs of those living in poverty. Constatine granted the Church the right to receive legacies as long as it agreed to support the poor.[5] The Church assumed the task of population pacification by providing for their needs and encouraging pious behavior. Throughout the Christian world Church fathers drew upon the Old and New Testaments to justify their view of poor people as divine. To be charitable was to practice the love of God by exercising mercy and kindness toward the less fortunate. According to the Gospel of Matthew (25:40) Christ is said to have identified himself with the destitute and commanded all his followers to care for them. It was through such works that one could achieve salvation. Michel Mollant tells us that the Christian concept of charity preceded the concept of the poor.[6]

By the Middle Ages poverty became an essential component of urban life. Yet it was not solely a judgment of one's economic position in society; it was part of a moral and social order. The Medieval Church preached that the destitute were in a state of grace—a position closer to God. Clerics would wash their feet and take vows of poverty as part of their holy orders. Holiness was inherent in poverty. Giovanni Francesco Bernardone (St. Francis of Assisi), founder of the Franciscan Order in 1209, came from a very wealthy home and dedicated his life to not only helping the poor but also to taking up this life way. But while he and other mendicant clerics preached peace and love many peasants who worked the land and were treated cruelly by large estate owners often rebelled against them.[7] It was sometimes difficult to portray the poor as pious.

Still St. Francis helped to establish an iconography of poverty based on peasantry and the peasant connection to nature. In fact it was this equation of the poor peasant with the land that fed those of the city, which helped to establish a place for the poor in some symbolic sacred and mystical order. But should the crop go bad, should the land refuse to yield its bounty, it was the peasant who was at hand to be blamed. Their mystical connection to nature was equally suspect.[8] In the 1300s the vast majority of peasants near major European cities worked very small parcels of land—often too little to provide any sort of security.[9] It was not unusual for those experiencing economic disaster to look to the city as a place for survival. Vagrancy grew into a significant problem. Peasants who could no longer command a livelihood from the land came into the city to sell their labor or to beg. According to Mollant, the thirteenth and fourteenth centuries began to see a proliferation of terms used to describe those living in poverty.[10] New terms for the poor also signaled a new awareness of its great diversity and accompanying misery. Throughout most of the Middle Ages only adjectives were used to describe types of poverty and states in which poor people existed. Poor was a condition. It wasn't until the late Middle Ages that "the poor" came to exist as a signifier—as a noun—a category encompassing different types of people. The urban poor were often portrayed more callously than their rural peasant brethren who stayed in their place and stood by their plows no matter what the circumstance. The poor that flooded the streets of medieval cities were met with distrust. They were for the most part strangers. It must be kept in mind that the number of poor always exceeded the more affluent businessmen, nobility, or clergy. During harsh times their ranks swelled tremendously.

Sharon Farmer's study of the poor in Paris and other medieval cities reveals that the destitute from the hinterlands came into the city to work as servants, prostitutes, and unskilled laborers. But they were also drawn to the city by the charity offered there and by the religious shrines that beckoned those with physical disabilities.[11] The urban poor often identified themselves as such. To be poor in the city was to be part of a large, marginalized, and disenfranchised group—a group with a particularly tainted identity. The laboring and nonlaboring poor comprised half of Paris' population in the late eleventh century. According to Farmer it was not unusual to see 10–20 percent of a city's population begging in the streets at any one time.

The portrayal of the poor in medieval times was often through the lens of Church artists. Artists depicted the poor as recipients of Christian charity. The Seven Corporal Works of Mercy, six of which were listed in the New Testament (Matthew 25:35), were means through which one could enter heaven by helping those who were disadvantaged: feeding the hungry, giving drink to the thirsty, clothing the naked, housing for the homeless, and the like. Wall paintings of the Works were popular throughout Europe from the fourteenth century until the Reformation. They adorned the interiors of many churches and served a heuristic function. In such paintings the poor beggars are usually portrayed as pathetic and miserable, with extended hand genuflecting or bowing to the charity of the patron. They are dressed in tattered clothing. Works by such artists as Giotto of St. Francis's life wherein he dismounts from his horse and assists a poor man by giving him his cloak (c. 1318) and an important work by Ambrogio Lorenzi in 1340 depicting St. Martin of Tours as a knight mounted on horseback cutting his cloak in two to clothe a pathetic, skinny, nearly naked beggar were exemplars of the Christian ideal of charity. In each case the saint's head is luminously haloed.

In his study of poverty in Early Modern Europe, Robert Jütte (1994) notes that artistic images of the poor provided "visual propaganda" that often promoted the moral programs of the Church.[12] "Around 1524 Nuremberg artist, Barthel Behaim, published his broadsheet *The Twelve Vagrants*. Its form as broadsheet, and its verbal format, in verse, enabled it to reach less literate audiences with its particular message, namely that sometimes the poor are victims of both internal and external forces, but that the presence of poverty is part of the divine order."[13] The imagery of the poor was forever connected to maintaining control and power over the masses. If people were not going hungry themselves they needed to feed those who were. This was part of the moral economy. The notion of exploitation or domination had little currency.

For certain, the great plagues of Europe in the fourteenth century killed vast numbers of the urban poor who were significantly more vulnerable to the spread of infection. Many who were crowded together in the worst parts of towns and cities succumbed to infection quickly, whereas nobles and clergy were disproportionately spared.[14] Part of this was due to the squalid living conditions—crowded, dank, and thin-walled shanties; but it was also their miserable diets. It was typically the urban poor who were hired to remove decaying corpses of their fallen brethren from the streets. In Renaissance Florence the afflicted poor were often quarantined, or consigned to pest houses and health boards that increased their death toll.[15]

Robust Bodies of the Poor

The perceived physicality of the poor separated them from elites. Their "robust" bodies that once worked the land appeared to be made for hard labor and for child production. But even though men who worked with their hands were considered masculine, their masculinity was frequently considered inferior to that of urbanites whose work was more cerebral.[16] They were far more animal-like. Men of lower order lacked the cultural qualities and refinements of nobles and clerics. Those who were crippled were inferior in terms of masculinity to those with robust bodies. According to Sharon Farmer, "A variety of thirteenth century texts elaborated the theme that disabled poor people were not only addicted to idleness but also despicable in their wretchedness. Some of their sufferings were punishments for sins."[17] Many authors of the thirteenth century developed themes of disabled beggars who preferred their disabilities to more robust bodies because they made so much money through them and did not have to work. These stories were often supplemented by plays and narratives of beggars pretending to be crippled so they could become rich.[18] Urban beggars were viewed as predatory.

Peasants and paupers were depicted as mentally inferior to the elites if not feeble-minded. Often peasants were depicted as filthy, subhuman, and comical "the reverse of the civilized and courtly" and frequently depicted as grotesque in appearance.[19] In art peasants were often likened to domestic animals or even represented by filth and excrement associated with their work.

Simultaneously, however, the peasants were recognized for their position of societal value because they fed the rest of the population. Their robust bodies were associated with agrarian upbringing and hard labor. The rural poor were people of the earth. But the poor who dwelled in cities were considered for the most part criminal and deviant. Their poverty was thought to be a matter of choice. This paradoxical position played itself out continuously particularly during the late Middle Ages. Peasants were seen as close to nature—close to God. This closeness to God was to disappear as they entered the city. The urban poor were for the most part undeserving and dangerous.

The Reformation and the Sinfulness of the Poor

The thirteenth century saw both the rise in importance of a new class of merchants and a growing ambivalence about the condition of the poor. With this came an increasing distrust of those poor who lived in the cities. Urban beggars and vagrants appeared to represent a threat to medieval order. As peasants continued to lose their land to enclosure and poor harvests, many headed to the city and joined the swelling ranks of the poor. During difficult times many resorted to thievery. In Florence throughout the 1400s most artisans endured bitter poverty. And in San Michele there were no fewer than 7,000 paupers.[20] By the late 1300s cities were experiencing what was labeled high vagrancy rates. Men, women, and children who could no longer return to pasturelands and who were now "masterless" flooded into the cities and towns.[21]

By the sixteenth century a "literature of roguery" gained popularity among the cultured classes. Vagrants were seen as becoming organized and posing a threat to urban life. According to historian A.L. Beier many were thought to be in league with Satan, and "Like witches they were thought to harm people with devilish practices."[22] There was a sinfulness associated with urban poverty that was not connected to their pastoral brethren.

The corruption of the Church and its espousal of the holiness of poverty were viewed as hypocritical. Artists such as Hieronymus Bosch (1450–1516) and Pieter Bruegel (the Elder (1525–1569) treated this phenomenon in their works. Drawing upon the teachings of the Church, such as the Corporal Works of Charity, they often presented a cynical and negative picture of the Church and its beggars. In their portraits the urban poor were often drunk, deformed, comical, or menacing. Clergy did not fare better.

If the Catholic Church was to beatify destitution, Protestants were often to speak of its sinfulness. It was primarily during the Reformation that poverty became a mark of shame.[23] Max Weber (1958[1904]) saw the emergence of industrial capitalism as intricately tied to the heightened sense of self that was associated with the Reformation.[24] The Protestant Ethic, which was central to Calvinism, frequently presented poverty and hard labor as a punishment from God. In Calvin's view predestination meant that God had already planned for some to achieve heaven and others to be condemned to hell. It was those who worked who would be closer to God and not those who remained idle. In the world of the flesh those who were chosen by God could be seen as materially rewarded for hard work and those who were not chosen were often condemned to lives of misery. To help the poor was to save their souls from the morally inferior lives, which they lived. But the poor needed to work. Thus this position was nearly the inverse of the Catholic Church's position. Both Calvin and Luther advocated toil and material restraint, not poverty, as the means of entering heaven. In so doing they tended to ignore many of the structural elements of economy and society that could devastate the economic well-being of the person. But they equally rejected the romantic notion that God loved the poor more than the rich, and that the poor were destined for heaven. Weber stresses the idea that teachings of the new religious sects were essential to the development of a capitalistic ethos.

In the fourteenth and fifteenth centuries the Church had already begun losing its control over the poor as serious revolts of peasants and other laborers broke out in villages and cities in Flanders, Germany, France, and Italy. The enclosure of lands forced tens of thousands of peasants from their homesteads into towns. By the 1500s Reformation leaders, particularly Luther, found themselves speaking out against these rebellions. Luther, who called for the complete extirpation of beggars from the Christian world called too for the mass slaughter of rebellious peasants and laborers in Germany during the Great Peasant Revolt.[25]

Throughout Europe vagrants and urban beggars were considered criminals. Laws requiring their detention and punishment were periodically adopted and stringently enforced. These were among the first attempts at eradicating their presence—at cleaning up the city streets. In England a statute of 1495 required that beggars and vagabonds be set in stocks for three days and fed nothing but a slim diet of bread and water. A second offense would result in doubling the penalty. But by the fifteenth century, punishment was even harsher. A.L. Beier (1985) notes: "Whipping was a sixteenth century development. The poor law of 1531 stipulated that vagrants be 'tied to the end of a cart naked and be beaten with whips' around the town or village, and until their bodies were bloody.[26] These appeared to be punishments without the benefit of trial, which frequently resulted in the vagrant leaving town. But there were far too many vagrants to make this an effective measure.

In France, the first real attempt to imprison the city poor and force them into labor was under the order of Marie de Medici. Three hospitals were built in Paris for this purpose. These were collectively referred to as *Hôpital de Panvres Emfermez* or Hospital for the Imprisonment of the Poor.

> In the autumn of 1611 a decree was announced forbidding begging in Paris and ordering all non-resident beggars to leave the city; local beggars were required to find work immediately, or failing this to present themselves, on the day appointed by the decree, at the *Place St. Germain*, to be taken to one of the hospices.[27]

Accordingly this decree so intimidated the beggars that only ninety-one of between eight and ten thousand beggars presented themselves to the authorities; but as economic conditions in the city worsened many more joined their ranks. Within six weeks eight hundred were confined and by 1616 twenty-two hundred wound up in these facilities.[28] Geremek reports that "Women who were caught begging were publicly flogged and had their heads shaved, while men were taken off to prison."[29] By the beginning of the eighteenth century, the vagrant and the beggar began to disappear from the city streets. They were incarcerated in prisons or workhouses along with street children, petty thieves, rapists, and lunatics.

Capitalism: Depicting the Urban Poor

The urban poor were not an invention of modern capitalism. They lived in cities long before the sixteenth century and long before the rise of nation states. But their appearance in cities was nearly always associated with deviance. While the Church was an instrument of their articulation and control, its loss of power and legitimacy loosened its authority over them and how they were defined. With the rise of modern capitalism goodness was articulated differently. Now it was measured more in terms of material well-being and comfort—order and stability. This was a notion that was promoted by the emerging bourgeoisie and the newly established Christian reforms. It was also connected to the rise of consumerism and was a fundamental tenant of the European Enlightenment. Poverty took many forms, but it was the idle disparate poor who represented urban social pathology. Idleness and all that was associated with it—such as disease, filth, and crime—was thought to be contagious. For this and many other reasons the poor needed to be further separated from those who found their closeness offensive. Poverty was messy and disorderly and force was needed to bring it under control. In the early sixteenth century in Italy, Germany, and France, and in the Iberian Peninsula, city magistrates watched their populations of vagrants and beggars rise along with crime rates. Segregation of these populations became essential.

Afraid of disease, criminality, and prostitution, the rich withdrew themselves from the old city centers and established themselves in new residential areas. Where that was impossible for military or other reasons the poor were driven to the periphery of the town or to special streets or quarters, which turned into ghettos.[30]

Thomas Hobbes, who saw large numbers of poor people as a threat to order and stability, had a profound influence on British social thought in the seventeenth century. His notion of the poor came from his interpretations of natural law and unique reading of the Bible. In the spirit of the Enlightenment, Hobbes attempted to drain religious mysticism from the sentimental depictions of the poor. Pillars of capitalistic promotion, including Adam Smith, David Ricardo, and Thomas Malthus did the same; but they often connected their views to a new secular Christian morality. Individual self-interest and the inherent laziness of the common rabble were seen as natural states; therefore, a need existed to exercise moral control over the working class to push them to be more productive. Capitalism and subsistence wages served a higher spiritual and social purpose—to bring order to society.

In 1798 Jeremey Bentham published his famous *Pauper Management Improved*. This was a companion piece to his *Panopticon*, later made famous by Michel Foucault. While the panopticon was a suggested model for prison surveillance he had also recommended its use in factories, asylums, and poor houses. But his plan for the "burdensome poor" called for a major social undertaking wherein a state chartered corporation would have total control over the destitute. It would begin by establishing five hundred work houses for one million people who would live and work in groups of two thousand. Each and every person would be under "constant" and "absolute" supervision. People would be forced into the house by denying them relief elsewhere.[31]

There is no paucity of grotesque images of the poor published during the early days of the industrial revolution. In the etchings of William Hogarth the idle poor are often grotesquely dehumanized. Their habits are grossly exaggerated. They flood the streets and are often depicted as drunks, lunatics, and deformed cripples. Historian Sarah Jordan points to the propaganda laid down by those wanting to better harness the energies of the working poor by denying them assistance of any form.[32] She quotes William Temple, a business spokesperson, in a 1758 essay on the importance of capitalistic production and the negative consequences of labor's "ungovernable appetites":

> If a labourer can procure by his high wages or plenty, all the necessities of life; and have afterwards a *residuum*, he would expend the same, either in gin, rum, brandy, or strong beer; luxurize on great heaps of fat beef or bacon, and eat perhaps till he spewed; and having gorged and gotten dead drunk, lie down like a pig and snore til he was fresh.[33]

Drunkenness and gluttony are seen as essential character traits of the nondeserving poor. There is an inability for them to control themselves. Where the respectable poor are clean, neat, and appreciative of what they have and attempt to emulate their superiors, the contemptible poor don't even try. In 1755 John Clayton authored *Friendly Advice to the Poor,* in which he admonishes them for being dirty and dressed in ragged clothing. He calls on them to be "tight and whole," and describes the loathsome poor as "distressed" and "wretched objects" with which "our streets swarm," but are so "familiarized to Filth and Rags, as renders them in a manner natural." He likens them to excrement."[34]

It is during the early period of market capitalism, punctuated by an increase in the power of the bourgeoisie and a corresponding diminishment of importance of religious power that the idle and working poor begin to be massed together as targets of condemnation, repression, exploitation, and control, or feeble attempts at "rehabilitation."

Where the poor had long been classified as either worthy or unworthy the distinction becomes rapidly degraded.[35] Poverty was no longer seen as a part of divine providence but rather as a matter of choice. And it was deemed to be much more of a choice among the urban poor than their rural counterparts who were still connected to the land and its sacredness. Poverty was a pathology of character—a pollution of the soul. One merely needed to enter the stench of the city to discover its rampant ugliness and danger. According to David Ward: "Poverty and other social ills were no longer viewed as providential and endemic, but rather as self-afflicted and reformable deviances."[36]

Gertrude Himmelfarb in her discussion of Frederick Eden's *History of the Laboring Classes in England* (1797), shows how Eden called for a redefinition of the poor that would exclude unemployed laborers or the physically able but destitute idle.[37] Eden makes the point that the poor are actually people who were once pastoral "slaves." Thus the poor are now free of their bondage to medieval estates. Before the introduction of manufacturing, he explains, there were no poor. It was capitalism that liberated peasants from their territorial lords and imbued them with personal freedom. To be poor was to be free—to be free to sell one's own labor and to consume. In this way the poor are now responsible for their own destinies. This revisionist history establishes a new conceptualization of the poor—one that became more generally accepted over time.

Turmoil, Revolution, and the Unsettled Crowd

The close of the eighteenth century was marked by the ascendancy of the bourgeoisie and the final days of the ancien régime. It was the French Revolution that signaled the rise of a prosperous middle class that was able to wrestle power from the throne of France. While it was the poor who rioted and fought the street battles, it was the bourgeoisie that eventually ascended to power and abolished the last remnants of feudalism establishing a more modern government that would serve the interest of business and commerce. However, the French Revolution was to set into motion a series of rebellions throughout continental Europe. While the revolution itself was initially seen as inspirational, the bloody chaos that followed close in its path was not. The condition of the poor changed very little.

Although the storming of the Bastille by hungry and unemployed workers was certainly an iconic event, the revolution was in no way limited to the city of Paris. The power of the poor to overtake and slaughter the fortress' guards was to mark the revolution's "savagery" and the recognition of the potential for violence inherent in the proletarian crowd. Peasants on the outskirts of cities also revolted, burned down the quarters' of their landlords, and demanded their freedom from feudal obligations. The world was changing. The poor gradually emerged as an inherently violent force needing to be recognized for its extremely destructive nature. The period between 1815 and 1848 was furthermore marked by counterrevolutions, repression, rioting, and despotic rule.[38]

It was from this mix of social turmoil, bloodshed, and revolution that Auguste Comte (1798–1857) proposed his new science of society—one that would apply the principles of positivism to secure a more stable and orderly world. It was his belief that the Enlightenment and the French Revolution were responsible for a state of anarchy and that intellectuals and writers had led France into a state of war with itself.[39] Comte worked closely with Claude-Henri de Saint-Simon (1760–1825)—a self-professed social engineer who envisioned a utopian society led by a council of industrialists and scientists.

Sociology was to help establish the tools for defining, counting, and measuring the poor. It was a social science that primarily focused on cities as the cauldrons of unrest and civil disorder. Comte's work was to set the stage for late nineteenth-century sociology including urban social survey movement. If one was to look closely at sociology and its emergence at this time it becomes evident that its central purpose was not remediation in the service of the poor but rather a tightening of control over them by societal elites.[40]

Still reformists remained in the wings. In some articulated liberal concerns about the urban poor that began spreading throughout art and literature at this time.

The Romance of Poverty and the Evil of Cities

As the industrial revolution sired smoke spewing factories across the face of Europe, and as the sanctified peasant was transformed into commodified urban labor, a change was occurring in artists' conceptions of the newly emerging order. While one might celebrate the victory of human spirit over the old aristocratic institutions of oppression and the hierarchical Church, one could also lament the passing of the naturalistic world shrouded in myth and spirituality. It was there in the mythical pristine pasturelands that the traveling minstrel serenaded the peasant girl beneath the brightest shining stars. It was there in the mountains, in the rain, that ancient spirits dwelled. But it was in the city where iron cages of reason, void of sentiment, awaited the arrival of the new rural migrant; where factories like Moloch devoured the working poor. These were the visions of bourgeois artists and writers. Sentimentalism for what was left behind in the great agrarian mist came to full flower in the eighteenth century, as did strong currents of antiurbanism.

This Sentimentalism was not merely a movement in literature prior to the birth of Romanticism, but rather an important philosophical worldview—perhaps even an ideology that was deeply rooted in anti-urbanism. As a response to an Age of Reason the Age of Sentiment was a powerful driving force in eighteenth century Victorian life.[41] Aside from placing value on feeling above thought, passion above reason, it was often manifest in an attack on bleak aspects of social change wrought by industrialization. While some of it called for a return to a more traditional social order, much of it was reformist in nature.

Barbara Benedict (1990) suggests that sentimental literature was aimed at poorly educated working-class audiences, both urban and rural, who consumed these writings that were often serialized and cheaply produced.[42] While much of this work never reached the standard that was to be set by George Crabbe, Charles Dickens, or Emily Barrett Browning, such writings were frequently accompanied by inexpensively reproduced engravings that depicted the tragedies of poverty. Dickens' illustrators set a standard for sentimental art that would accompany his tales and those of other writers of this era. John Millais (1829–1896), Charles Allston Collins (1828–1873), Marcus Stone (1840–1921), Samuel Luke Fildes (1843–1927), all had their hand in illustrating his works with bleak portraits of the poor and their living quarters.[43] But these were mostly portraits of the urban poor as victims. Fildes 1891 sketch, *The Homeless and Hungry,* that appeared in *The Graphic* was a case in point. Here adults and small children dressed in ragged clothes wait huddled in the cold street outside a workhouse awaiting the possibility of tickets to spend the night there. However, his engraving was done to run alongside a story calling for the adoption of the Houseless Poor Act.

If Sentimentalism was to find its expression in descriptions of human degradation in urban life, it was also to find its way into the sense of what people had lost. Crabbe's and Wordsworth's poetry did much to remind their audience of this. The birth of this *new* romantic pastoral was a reaction to the spreading of industrial cities and the values that went along with it. The Romantic Movement in the arts often contributed to the emerging perception that cities were the generators of ruthlessness and poverty—wiping out nature along the way.

Blake's walk through London's "chartered streets" wherein he found on "every face" he met "Marks of weakness, marks of woe," sharply contrasted with his view of nature and its inherent spirituality promoted in his "Introduction to his Songs of Innocence," or in his lyrical poem "The Shepherd," celebrating the rustic's "sweet lot." In the early work of Wordsworth we find pastoral sentimentality at least equal to this. However, it is Wordsworth poem to the glory of London, "Composed on Westminster Bridge" that he displays both a versatility as a poet and the makings of a fine urban propagandist. While many of Wordsworth's poems are populated by vagrants and paupers, these characters appear to be mostly alone. Their solitary position seems to give them a dignity that they don't have crowded together. Unlike Dickens' many characters, they do not seem dangerous or threatening. In no way are they revolutionary. They represent something lost.[44]

Capitalism was represented by greed as well as machine and coal-driven industries that spewed filth into the air forcing the poor to find shelter in the most miserable of abodes. These artists and writers identified problems of the industrial cities with the problems of the poor who lived there—poor sanitation and inadequate housing; danger and crime; prostitution and vulgarity. Slums came to represent the character of the poor themselves—their filth, their deviance, and their broken down spirit. It is significant that the urban poor were often considered more "polluted" than their rural counterparts whose poverty was perceived as much more natural—perhaps more undeserved. The urban poor were often relegated to a distinct area of the city or close to its outskirts. Housing in these bleakest quarters of the city was often makeshift, constructed of the cheapest of materials. Unlike the better housing in other sections of the city, this construction was ramshackle, built close to factories, and often contaminated by their smoke and waste. It was overcrowded and a breeding ground for disease.

It would be safe to conclude that the Age of Sentimentalism as well as Romanticism saw "respectable" rural poverty retreat into the naturalistic landscape—a much healthier place to live and raise one's family. There was nothing wrong with it there. The country was perceived to be a safer place: one more spiritual and respectful of order and tradition. While people might be poor, they were not defeated; they were not deranged or damaged. They were not dangerous.

This same sentimentalism is evident in the work of social thinkers and philosophers ranging from Ferdinand Tönnies to Oswald Spengler, but is more sophisticatedly developed in the work of sociologist Emile Durkheim.[45] It is contained in the urban-rural dichotomy established not only as a powerful Western aesthetic, but also a powerful theme in social theory.[46] Raymond Williams' *The Country and the City* provides his reader with a most remarkable understanding of how this particular aesthetic came to influence important social thought of this time.[47]

But if sentimentalism and a romance with nature were to characterize one modernist tendency in the arts, the infatuation with science, reason, and production was to exemplify another.

Modernity, Capitalism, and the Revolutionary Poor

While many social critics looked either contemptuously at the working class or saw them primarily as victims, Karl Marx saw within this group the seeds of revolutionary change. His view of the urban poor is far less sentimental, far less pessimistic than most of his contemporaries. His social science is an aesthetic realism associated with modernism.[48] He embraces the city as a realistic venue for social change, especially in his *The Eighteenth Brumaire of Louis Bonaparte*.[49] In his writings he recognizes that the true challenge of the working poor is to define themselves. Marx understands that to control the definition of the working class is to control the working class itself. *The Communist Manifesto*, more than an epistle of revolution, is a call upon the laboring poor to both invent themselves and recognize their own potential for doing this.

By calling into question the foundational elements of capitalism, and how it defined the world, Marx encouraged his readers to imagine a place and time without it. Inherent in Marx's notion of dialectical materialism is a *radical personalism*. This was later articulated in the works of Frankfurt school theorists, but most importantly in the writings of Erich Fromm.[50] Self-imagining was the very basis for all revolution.

Unlike sentimentalists of his era, Marx rejected the pastoral romance. It was his firm belief that capitalism had ushered in wondrous industrial and technological achievements and had opened up space for perpetual revolution. Such change could only take place in the cities where crowds congregated and new ideas were generated. He saw in industrial technology a massive materialistic reorganization. He recognized that all that had come before was now rapidly disappearing and that this would be a perpetual process. He celebrated this.

Marshal Berman has spoken of the power of industrial capitalism to change not only the urban landscape, but also perceptions of that landscape through both planning technologies and media.[51] Like Neil Smith[52] and Mike Davis,[53] Berman has seen in the process of urban gentrification a counterrevolutionary spirit aimed at population pacification and control. Programs of population relocation, commencing with grand modernist schemes of urban renewal and revitalization, often granted cover to more sinister programs aimed at erasing the poor from urban life. Baron Haussmann's plan for the renovation of Paris, commissioned by Napoleon III, was a case in point. There can be no coincidence that this project came on the heels of the 1848 commune battles of Paris that saw the poor and working class confront despotic authority. It was with this plan that the poor were basically chased from the center of the city, their homes destroyed and replaced by wide boulevards strung with new cafes all aimed at bourgeois consumerist culture. In *All That Is Solid Melts into Air,* Berman draws from Baudelaire's "Eyes of the Poor," to depict the ghost-like quality of the urban poor who gaze hauntingly into a café window disturbing the sensibilities of the middle-class couple within. The poor disappeared into the night of a modern metropolis. It is not a far jump from Haussmann to Robert Moses.

Throughout the twentieth century governments worked to segregate populations of working-class peoples—particularly by race. Slum clearance projects aim at eradicating "urban blight" and laws to encourage the white working class to move from the cities were vigorously adopted. Public housing initiatives further isolated the distressed poor—chased them from the city's center. By the close of the twentieth century global capital was determining the life or death of communities—the life and death of cities.

Modern Media: Race and Poverty

But it was far worse than just eliminating the poor. It was portraying the urban poor as a cancer that needed to be cut from an otherwise pristine body. This portrayal of the poor as greedy, diseased, dirty, ravenous, criminal, and insane was used to develop programs aimed at their removal from the urban landscape. Should the poor be eliminated then the city could once again be generous, healthy, clean, restrained, orderly, and rational. Perhaps it could, once again, be white. It is not coincidental that the jails of bourgeois cities now bulge with the poor inmates and particularly poor people of color.[54]

There is very little doubt that these characteristics associated with the poor are features many come to associate with the offensiveness of urban life. Some of these stereotypical urban attributes are racist codes. Some come out of this false urban-rural binary—out of a false binary of undeserving and the deserving poor. In their research on media portrayals of the poor, Rosalee A. Clawson and Rakuya Trice found that large numbers of Americans view poor people as exhibiting undesirable characteristics that violate mainstream ideals. Much of this understanding comes through the mass media.[55] The media frequently describes the underclass in behavioral categories such as criminals, alcoholics, and drug addicts.[56] Clawson and Trice's media analysis concludes that while African Americans make up less than one-third of America's poor, the mass media lead people to believe that two out of every three poor people are black.[57] The majority of Americans believe the claim that the poor "lack effort" and have "loose morals."[58]

In his 2004 survey as to how race enters into this conceptualization Martin Gilens found that of those white respondents (to his questionnaire) who accurately believed that the majority of welfare recipients in the country were white, fully 50 percent thought their being on welfare was because of "circumstances beyond their control." Of those white respondents who believe welfare recipients were primarily black only 26 percent believe it was "circumstances beyond their control"; 63 percent of these white respondents attributed welfare to "a lack of effort."[59] Even the portrayals of the poor in the media in the 1990s grossly exaggerated the proportion of people who were black. Well over 60 percent of photographs illustrating stories that discussed poverty in magazines such as *Newsweek, Time, and U.S. News and World Report* were of African Americans who represent only 26 percent of the poor.[60] Beyond this black faces were disproportionately associated with unsympathetic welfare stories (welfare cheating and corruption) whereas when whites were shown they were related to more sympathetic stories regarding poverty and welfare.

In 2000 Clawson and Trice found that magazine descriptions of poverty showed it to be almost exclusively urban. Ninety-six percent of the poor were shown as living in urban areas. Fewer than 30 percent of the poor portrayed in media were shown to be working people where in reality over 50 percent of the poor work full- or part-time jobs.[61] If this isn't enough, in 2002 Clawson and Trice discovered that such race coding has made its way from the popular media into textbooks. In their study of college level economic texts they found that such materials overrepresent the proportion of black people living in poverty while white people are celebrated for their achievements.[62]

Antiurbanism and the Depiction of the City Poor

The danger of urban poverty has always been recognized; less so its rural counterpart. In art and literature in the West, it was often represented through images of chaos, violence, and debauchery—disorder that threatened the social balance of civilization itself. In 1895 Gustav LeBon portrayed the urban mass as primitive, impulsive, and "too mobile to be moral."[63] He rhetorically asks why this is so and responds: "Simply because our savage, destructive instincts are the inheritance left dormant in all of us from primitive ages."[64] For Le Bon the city crowd represents this primitivism. It becomes the place for a loosening of inhibitions and social bonds. In some ways the city crowd represents that which is repressed beyond its borders—otherness of all sorts.

Antiurbanism reflects this fear of *otherness*. Cities are portrayed as storehouses of people who are "unlike the rest of us." These people are far less patriotic, nationalistic, and far more ruthless than their rural counterparts. They are less settled and more detached. They are often "heartless," and far more "calculating." The urban poor are criminals, cheats, and frauds. They are primitives, not simply people whose lives are gone wrong. Like their more affluent urban neighbors who run the banks and control the media, they are untrustworthy. Cities are the harbingers of these dangerous strangers—of immigrants, gay people, welfare cheats, and accountants.

But what is it that produces this unwholesomeness? Traditionally cities have been defined by sociologists such as Louis Wirth by their density, population size, and heterogeneity that are said to result in "an urban way of life."[65] The anonymity that comes as a consequence of these variables generates alienated strangers and what sociologists once liked to call "social disorganization." But in actuality the city constitutes a projection of the unacceptable—the chaos, violence, and the turmoil of a self that has been deeply repressed. In some ways, Le Bon touched upon that.

The city of the poor, just like the city of the rich, is a construct of the modernist imagination. The false binary of good and bad, deserving and undeserving constitute a simplistic way of organizing reality. It is safe. Like all false binaries, however, these discrete categories cannot be defined except through power. Psychoanalytically the construction of urban poor constitutes a Klienian "split." Here the world is divided between good and evil (deserving and undeserving) so that the self is justified in its rage. But this rage against the urban poor comes from a deep emotional insecurity that is the product of social instability and constant change. It is a projection outward of the dangers housed within.

Melanie Klein was able to capture in her psychodynamic model the fragmentation inherent in all modernist change. The splitting she articulates is a means of protecting the vulnerable self from fear of otherness and disorder. Here the urban poor come to represent the evil of the city itself: its violence, disorder, and decay. In imagining the urban poor in this fashion one is drawn into an illusion that the evil of the city is its poor.[66] (Segal 1974). The paranoid position she articulates in her work underlies the false binary that establishes the foundation of antiurbanism.

Conclusion

In *Icon and Idea* Herbert Read drew from the work of philosophers Conrad Fiedler and Cassirer to present the notion that art has been essential in the development of human consciousness.[67] Once created, posits Read, images are ever-lasting. We come upon ideas only in contemplation of images that we can draw upon. The enigmatic that is encountered in the material world is made sense of by and through imagination.

Likewise Robert Nisbet identifies certain tropes that run through art, literature, and sociology. One of these is the urban-rural binary that forms the basis for most discussions of antiurbanism.[68] The creation of sociology is predicated on the social scientific articulation of the same imagery of the urban poor we find in nineteenth-century poetry and drama. Whether Engels or Blake we are presented with binary images that allow us to draw upon an iconology that links history to contemporaneousness.

Social history exposes its important foundational elements that have entered into our construction of the urban poor. Such a construction is not merely one of imagination, but something that is fabricated from the material world. But it is the imagery of the poor that has evolved with time—not poverty itself. It is both sacred and profane. It is natural and unnatural. Poverty is the basis of war and death; it is the basis of simplicity and order. It is the foundation of chaos and control. In her 1969 essay on the naturalization of poverty Ruth Smith borrows a concept from Baudrillard that posits the importance of poverty's association with *good nature* and *bad nature*.[69] Good nature, according to the Enlightenment, is nature that can be harnessed by modern society; bad nature is its uncontrollability. The division of the poor into camps of deserving and the undeserving reflects this division in nature—the docile and the deviant. The various narratives and portrayals of the poverty and the poor seem primarily designed to come to terms with its enigmatic nature. That it exists is a testimony to science's inability to conquer and control it. It is denied, challenged, and attacked. Wars have been declared on it but it does not appear to be going away. It was with us before capitalism and remains a central component of the modern neoliberal capitalistic order. It gives life to that order.

Whenever liberal social scientists speak of poverty it is not so much in contrast to wealth, but related to the disorder it breathes, or its inherent ugliness—the danger it poses to an orderly society or its incompatibility with modern life. The most powerful forces have been recruited to deal with it: the military, the state, religion, technology, capitalism itself. It is frequently reconceptualized out of existence but too often sneaks back and visits us in the night. Imagining poverty doesn't take a lot for many of us. It is part of the same universal binary that is inherent in urbanism itself. It's what keeps the engines of capitalism well oiled.

Notes

1. Michael Mollant, *The Poor in the Middle Ages* (New Haven, CT: Yale University Press, 1986) 33.
2. Ibid., 16.
3. Ibid., 15–16.
4. Ibid., 18.
5. Ibid., 20.
6. Ibid., 21.
7. Ibid., 82.
8. Carolyn Merchant, *The Death of Nature: Women, Ecology and the Scientific Revolution* (San Francisco: Harper and Row, 1980).
9. Catharina Lis and Hugo Soly, *Poverty and Capitalism in Pre-Industrial Europe* (Atlantic Highlands, NJ: Humanities Press, 1979).

10. Mollant, *The Poor in the Middle Ages*, 2.
11. Sharon Farmer, *Surviving Poverty in Medieval Paris* (Ithaca, NY: Cornell University Press, 2002) 32.
12. Robert Jütte, *Poverty and Deviance in Early Modern Europe* (Cambridge: Cambridge University Press, 1994), 14.
13. Ibid., 19.
14. Mollant, *The Poor in the Middle Ages,* 193.
15. Ann G. Carmichael, *Plague and Poor in Renaissance Florence* (Oxford: Oxford University Press, 1986).
16. Ibid., 47.
17. Farmer, *Surviving Poverty in Medieval Paris,* 61.
18. Ibid., 64–65.
19. Paul Freedman, *Images of the Medieval Peasant* (Stanford, CA: Stanford University Press, 1999) 157.
20. Lis and Soly, *Poverty and Capitalism in Pre-Industrial Europe,* 17.
21. A.L. Beier, *Masterless Men the Vagrancy Problem in England 1560–1640* (London: Methuen,1985).
22. Ibid., 8.
23. Mollant, *The Poor in the Middle Ages,* 1.
24. Max Weber, *The Protestant Ethic and the Spirit of Capitalism* (New York: Charles Scribner's, 1958 [1904]).
25. Bronislaw Geremek, *Poverty a History* (Oxford: Blackwell, 1994), 180.
26. Beier, *Masterless Men the Vagrancy Problem in England 1560–1640,* 159.
27. Geremek, *Poverty a History,* 221.
28. Ibid., 221.
29. Ibid., 221.
30. Lis and Soly, *Poverty and Capitalism in Pre-Industrial Europe,* 79–81.
31. Gertrude Himmelfarb, *The Idea of Poverty: England in the Early Industrial Age* (New York: Alfred A. Knopf, 1984), 79.
32. Sarah Jordan, "From Grotesque Bodies to Useful Hands: Idleness, Industry, and the Laboring Class." *Eighteenth Century Life* 25 (2001): 62–79.
33. Ibid., 64.
34. Ibid., 67.
35. David Ward, "The Victorian Slum: An Enduring Myth?" *Annals of the Association of American Geographers* 66 (2) (1976): 323.
36. Ibid., 332.
37. Himmelfarb, *The Idea of Poverty: England in the Early Industrial Age,* 75–76.
38. J.L. Talmon, *Romanticism and Revolt: Europe 1815–1848,* (New York: Harcourt, Brace & World, 1967) 32.
39. Roger Salerno, *Beyond the Enlightenment* (Westport, CT: Praeger, 2004), 18.
40. Herman Schwendinger and Julia R. Schwendinger, *The Sociologists of the Chair* (New York: Basic Books, 1974), 75–80.
41. Louis Bredvold, *The Natural History of Sensibility* (Detroit: Wayne State University Press, 1962), 50.
42. Barbara M. Benedict, "Literary Miscellanies: The Cultural Meditation of Fragmented Feeling," *English Literary History* 57 (2) (1990): 407–430.
43. Frederic George Kitton, *Dickens and His Illustrators* (Honolulu, HA: University of the Pacific, 2004).
44. Gary Harrison, *Wordsworth's Vagrant Muse* (Detroit: Wayne State University Press, 1994).
45. Emile Durkheim, *The Division of Labor in Society,* (New York: Free Press, 1984 [1893]).
46. Robert Nisbet, *Sociology as an Art Form* (New York: Oxford University Press, 1976).
47. Raymond Williams, *The Country and the City* (New York: Oxford University Press, 1973).

48. Eugene Lunn, *Marxism and Modernism* (Berkeley, CA: University of California Press, 1982).
49. Karl Marx, *The Eighteenth Brumaire of Louis Bonaparte* (New York: International Publishers, 1987 [1852]).
50. Erich Fromm, *Marx's Concept of Man* (New York: Frederick Unger, 1961).
51. Marshall Berman, *All That Is Solid Melts into Air* (New York: Penguin Books, 1982).
52. Neil Smith, The Urban Frontier: Gentrification and the Revanchist City (New York: Routledge, 1996).
53. Mike Davis, *City of Quartz* (New York: Vintage Books, 1992).
54. Jeffrey Reiman, *The Rich Get Richer and the Poor Get Prison*, 8th ed. (Boston, MA: Allyn & Bacon, 2006).
55. Rosalee A. Clawson and Rakuya Trice, "Poverty as We Know It: Media Portrayals of the Poor," *Public Opinion Quarterly* 64 (1) (2000): 53–64.
56. Clawson and Trice, "Poverty as We Know It: Media Portrayals of the Poor," 54.
57. Rosalee A. Clawson and Rakuya Trice, "Poor People, Black Faces: The Portrayal of Poverty in Economics Textbooks." *Journal of Black Studies* 32 (3) (2002): 54.
58. James R. Klueger and Eliot Smith, *Beliefs about Inequality* (New York: Aldine De Gruyter, 1986), 79.
59. Martin Gilens, "Race Coding and White Opposition to Welfare." *American Political Science Review* 90 (3) (1996): 593–604.
60. Ibid.
61. Clawson and Trice, "Poverty as We Know It: Media Portrayals of the Poor," 60.
62. Clawson and Trice, "Poor People, Black Faces: The Portrayal of Poverty in Economics Textbooks," 352–361.
63. Gustav LaBon, *The Crowd: The Study of the Popular Mind* (Mineola, NY: Dover, 2002 [1895]).
64. Ibid., 24.
65. Louis Wirth, "Urbanism as a Way of Life." *American Journal of Sociology* (44) (1) (1938): 1–24.
66. Hanna Segal, *Introduction to the Work of Melanie Klein* (New York: Basic Books, 1974).
67. Herbert Reid, Icon and Idea: The Function of Art in the Development of Human Consciousness (Cambridge, MA: Harvard University Press, 1955).
68. Robert Nisbet, *Sociology as an Art Form* (New York: Oxford University Press, 1976).
69. Neil Smith, The Urban Frontier: Gentrification and the Revanchist City (New York: Routledge, 1996).

CHAPTER EIGHT

*Americans, Urbanism, and Sprawl: An Exploration of Living Preferences**

Emily Talen

One of the most persistent cultural truisms about Americans is that most of them have a general dislike for cities and urban life. The common perception is that Americans, still under the spell of Jeffersonian agrarianism, equate urbanity with immorality. Jefferson's view that cities are "pestilential to the morals, the health and the liberties of men" and constitute "a malignant social form . . . a cancer or a tumor" has been used repeatedly to expound upon this view and diagnose the sad state of contemporary American landscapes. In large part the disdain is tied into the endless pursuit of the American Dream, and classic studies of suburbia like Kenneth Jackson's *Crabgrass Frontier* and Robert Fishman's *Bourgeois Utopias* have investigated both the causes and effects of this enduring quest.[1]

Yet there is a growing recognition that American attitudes toward urbanism are complex and variable. In some way, the recurrent goal that cities be compact, walkable, and diverse, as pronounced in the smart growth and sustainable development movements, is based on this potential. It essentially *requires* that the American conception of cities be more nuanced. In other words, it is necessary to allow flexibility in American attitudes if we are going to continue pushing a model of urban landscape reform that goes against the grain of current settlement pattern. Some will interpret this as "wishful thinking," but it is not unreasonable to suspect that popular support for smart growth models is more than antidevelopment. Smart growth requires that at least part of the population be willing to live more urbanistically.

An enduring problem for smart growthers and sustainable city advocates is how to successfully articulate the proper relation between the city and nature, between the urban and the rural. Anselm Strauss identified this as the essential dichotomy of American life and thought. While the division has been described as "potent," it does not necessarily suggest antiurbanism. It does, however, suggest a conflictedness in American and Western European culture, described by Tony Hiss as "mental baggage" that originated as early as the industrial revolution, and that continues to bewilder us by equating progress with environmental degradation. Americans have always had a hard time finding the right balance, claiming as Lewis Mumford did that on the one hand "urban and rural, city and country, are one thing, not two things," but on the other hand, that the city is an "artificial environment that enhances the dominance of man and encourages an illusion of complete independence from nature." Translate these discordant views into a program for rectifying the problem of human settlement, and there are bound to be tensions and conflicts.[2]

Some scholars, notably Fishman, argue that American antiurbanism is a "persistent misunderstanding," and that in fact "in no other society since the European Middle Ages have cities played such a formative role in creating the national economy and culture" (p. 6). It could also be argued that, two hundred years after Jefferson, accounts of urban vileness are more about the condition in which Americans have allowed cities to devolve into than a denouncement of the viability of cities per se. Further, it is possible to see a rift between what ordinary citizens think about cities, and the bulk of writing about cities from the American intelligentsia. Historically, there were plenty of boosters, orators, ministers, and common folk who spoke passionately for cities. Thus, to some extent, the great fear and anxiety of urbanism was born from the American intellectual, not the common urban dweller.[3]

This chapter is aimed at furthering our understanding of American living preferences by looking at just one indication of how places are valued—visual preference. It does not seek to scientifically verify, in the manner of housing market studies, that Americans will be more likely to move to one type of living environment over another. Nor is its intent to refine the definition of concepts like "urbanism" and "sprawl" in more precise terms. Instead, the survey results reported in this chapter should be interpreted as an initial exploration, a quantitative but largely exploratory investigation of people's impressions of different kinds of places, as represented by different kinds of images.

The survey results are thus intended to add to the body of empirical research about American views on cities, contrasting preferences about different forms of cities with views about the urban antithesis—sprawl. The focus is on understanding how the (mostly) negative conditions of sprawl and the positive and negative conditions of urban living are generally perceived. Pleasant suburban conditions are excluded—a survey of relatively innocuous suburban conditions would be unlikely to reveal anything new about the underlying complexity of preference.

One of the main contributions of the survey is that it uses a unique empirical approach—a nationally stratified, random sample survey of visual preferences. This methodology was made possible by using a national panel of respondents who use Web TV technology to periodically respond to surveys. These generally take the form of marketing studies, ranging from preferences on cars, food, clothing, and now, living environments. Previously, researchers have mostly had to rely on "opportunity sampling" in visual preference surveys, an approach that is not randomized and usually consists of very small samples. With the new technology used in this study, it is no longer difficult to accomplish randomized responses to visual preferences.

The chapter is meant to inform the debate on American settlement patterns and what to do about them: the idea that we need to promote places that are compact, walkable, mixed in use, and pay greater attention to public space. In order to better understand the propensity for Americans to live more urbanistically, it is important to evaluate American attitudes about a range of living conditions.

American Preferences

Broadly speaking, it is safe to state that Americans prefer single-family detached to attached and multifamily housing, and green space over nongreen urban space. According to Shlay, this has been so well documented that the desire has become virtually synonymous with the "American Dream." Yet the preferences are not as straightforward as it might at first seem. As Reid Ewing put it, suburbanites prefer low density housing, but they "could do well without the rest of the suburban package." Despite the preference for low density housing, there is evidence that many Americans prefer self-contained, mixed use settlement types over single-use suburbs. Empirical studies have provided evidence that Americans value small towns over suburbs, and rank suburbs well below other single-family environments such as villages.[4]

At least some percentage of the American populace is willing to make adjustments to their current living environments. One study of the trade-off between pedestrian proximity to urban amenities and large lot size showed that 30 percent were willing to make the trade-off. Another study documented that 38 percent of respondents in one suburban neighborhood were willing to live in older, more urban neighborhoods close to downtown if the associated negatives (crime, poor schools) could be avoided. Related to this, in a study by Shlay, support for suburban locations and physically homogenous neighborhoods was found to be less important than the ability to control one's environment.[5]

One indication of the complexity of American living preferences is the sheer number of preferences Americans are able to identify. Brower reviewed 36 studies of residential preference and gleaned 33 separate qualities considered to be the most important attributes of "good" neighborhoods. Qualities ranged from "good maintenance" (the most frequently cited quality), to racial and ethnic homogeneity. The complexity may indicate a certain level of ambiguity. Studies have shown that unmarried women and employed married women have conflicting wishes that can't be reconciled in a single ideal neighborhood: the proximity to services characteristic of central cities, as well as the close social ties they presume exist in suburban environments. Shlay argues that, when given a range of options, preferences are in fact flexible.[6]

When it comes to sprawl, the main confliction in American preferences is straightforward: Americans like suburban living but they hate the consequences. They like living on large lots near big cities, but are unhappy with the resulting commute times and environmental degradation. Studies seem to reveal a pattern of Americans wanting things all ways, unwilling to make the necessary changes that would be required to have a less conflicted view about cities, suburbs, and living environments.

There are numerous examples of this phenomenon. A survey conducted by American Lives found that residents preferred walkable, mixed use, pedestrian oriented town design, but a majority said that they were unlikely to give up their cars. A recent poll of 1,003 Americans conducted by Gallup for the annual Environment/Earth Day revealed "a disconnect between concern for the environment and support for further measures to protect the environment." While 47 percent had a negative view of environmental conditions in the United States, up from 38 percent the year before, fewer numbers than the previous year were in favor of taking aggressive action. Another poll conducted in November, 2002 in the San Francisco Bay area showed that, on the one hand, commuters saw regional congestion as a major problem, but on the other, were unwilling to change their travel behavior. One respondent was later quoted as saying "I wouldn't say I'm satisfied with my commute, I would say I'm sort of fatalistically accepting."[7]

A number of authors have looked into the fluctuations and seeming inconsistencies in American values. David Myers, in his book *The American Paradox: Spiritual Hunger in an Age of Plenty* examined 40 years' worth of opinion surveys and concluded that Americans are not happier nor more materially satisfied today than 40 years ago, despite a significant increase in income and consumption levels. What is interesting is that between 1966 and 1994 the percentage of new homes with more than 2,000 square feet increased from 22 percent to 47 percent and yet, in roughly this same time period, American satisfaction with their financial situation dropped from 42 percent to 30 percent.[8]

An obvious question is whether it is possible to explain these variations in living preferences. A number of general observations based on a review of studies of residential preferences can be made. First, traditional two-parent families with children and full-time homemakers show the greatest preference for suburban living . Social interrelationships and psychological feelings associated with a particular place—concepts like attachment and sense of community—are sometimes associated with stronger feelings toward neighborhood. O'Brien and Ayidiya summarize the phenomenon as "the extent to which individuals feel a part of a neighborhood community serves as a critical intervening variable through which other objective and perceived neighborhood conditions impact on their subjective quality of life" (p. 21).[9]

As a general rule, people tend to feel most favorable toward the places they are currently living in. Predictably, people living in suburbs rate large lots and green open spaces more importantly than those who live in more urbanized environments. Similarly, residents in more central urban locations value accessibility to services over large lots, and support for proximity to community facilities is higher among apartment and condo residents who are likely to have greater existing proximity than low density suburban residents. In a study by Rothblatt and Garr, it was reported that residents of high density housing preferred inlying neighborhoods and higher density housing over outlying low density suburbs. Some research indicates that women prefer central cities to suburbs because of proximity to amenities in more urban locations, but the opposite conclusion has also been reported.[10]

Finally, one recent study evaluated the likelihood that wealthy suburban residents in the United States would endorse compact living, or "traditional urbanism." The study investigated whether upscale suburban residents could feel positive about living environments that were not low density suburban areas, but were compact and pedestrian oriented. The goal of the study was to investigate what dimensions of suburban preference appear to be changeable. It was hypothesized that positive feelings about traditional urbanism would correlate with (1) low levels of neighborhood attachment; (2) low levels of satisfaction with suburban developments; and (3) high awareness of the social and environmental liabilities of sprawl. It was further hypothesized that women, residents with children, high car-dependency, and previous experience living in a more compact urban setting would correlate positively with traditional urban environments.[11]

The survey of 185 randomly selected residents in an upscale suburb of Dallas revealed that preference for traditional urbanism was not correlated with attachment, nor was it associated with the degree to which suburban residents thought of suburbia as problematic in environmental or social terms. On the other hand, low levels of satisfaction with suburban living were associated with greater acceptance of traditional urbanism. In terms of resident characteristics, the variable with the most significant association with acceptance of traditional urbanism was noncommuting travel time. Other characteristics like gender, age, presence of children, work status, and background were not correlated.

Inconsistency and variability of preference, as well as inconclusiveness about what might be used as explanatory variables, may all simply point to the fact that American preferences are flexible and adapting or that many Americans are not as sure of what they want as we might think. This is good news for those involved in trying to change the American pattern of settlement, with the goals of limiting sprawl, protecting against environmental degradation, and making urban environments more compact.

Dependent Variables

The goal of this study is to add to the empirical body of knowledge about the living preferences of Americans by making use of a randomized visual survey. The survey approach was limited to exploratory issues only, to yield some general impressions about how Americans perceive certain types of urban places. The survey, it is hoped, contributes to our understanding of how Americans feel about cities, and, in addition, how they feel about urban places when held up against suburban sprawl. I especially wanted to capitalize on the graphic portrayal of these different forms of living environments.

It was necessary, then, to try to sort out what aspects of urbanism and sprawl were of particular interest, and find a way to represent these qualities in a set of visual images that would serve as the dependent variables in the survey. The main concern was to find a set of images that could tease out differences in living preferences in such a way that something could be said about the specific characteristics of what is liked or disliked. There was no need to include obvious examples of urban blight, such as images of vacant buildings or environmentally degraded landscapes. Instead, the goal was to look for examples of (1) characteristics of urbanism in different forms and contexts; and (2) different characteristics of sprawl. In this regard, it was not the goal of this survey to represent "sprawl" and "urbanism" as objective constructs. On the contrary, I wanted to test people's preferences toward pervasive characteristics that are, in some sense, stereotypical.

My approach for determining what these characteristics of urbanism and sprawl might be was to first review the literature on the qualitative aspects of urbanism. For example, Christopher Alexander's "timeless way of building" includes principles such as a human scaled design, compactness, pedestrianism, and civic-mindedness (in the form of quality public spaces). The concepts of diversity, compactness, and the importance of the public realm is pervasive, and reflected in the writing of such notables as Jane Jacobs, Lewis Mumford, and Kevin Lynch. William Whyte's views about the role of public space and pedestrianism; and Jonathan Barnett's analysis of the "fractured metropolis" were also relevant. New Urbanist literature was consulted for its representations of the qualities that contribute to good urban form.[12]

In order to put these ideas into an appropriate pictoral representation, I reviewed the literature on visualization of landscapes and urban environments. There has been important scholarship on visual evaluation of the built environment, especially Nasar and Nelessen. Nelessen's use of the "Visual Preference Survey," although not scientific in the sense of being randomized, has been used across the country to garner support for traditional neighborhood design. Nasar's more rigorous investigations of environmental preference have found that preferences for landscape elements like edges and nodes vary significantly with context, and that it is possible to elicit shared meanings among a populace about factors like likability. These empirical works follow in the tradition of visual preference studies that have long been a part of environment and behavior research. Important works include a study by Peterson entitled "Measuring Visual Preferences of Residential Neighborhoods" in which qualities like "beauty" and "closeness to nature" were rated based on a series of photographs. Kaplan and Wendt tested the preference for natural scenes using 56 slides, and found that complexity was an important determinant of preference, and that natural scenes were preferred to urban ones. Interestingly, both urban and natural scenes were preferred to suburban ones. More recently, Brower's survey of neighborhood preferences used descriptions of different types of environments to illicit in respondents a visual image of the kinds of places they valued most.[13]

Literature that focuses on the visual experience of urban environments is also relevant. These include Kevin Lynch's *The Image of the City*, Amos Rapoport's *The Meaning of the Built Environment*, Allan Jacobs' *Looking at Cities*, and John Jakle's *The Visual Elements of Landscape*. A more recent investigation of urban imagery and its meaning, the "visually based narratives about the potential of places," was presented in *Imaging the City*, edited by Lawrence Vale and Sam Bass Warner, Jr. Collectively, this literature has been valuable in promoting a deeper understanding of how people experience and interpret their surroundings.[14]

Visual preference studies and theoretical work on image evaluation and meaning was useful for selecting dependent variables—images—and for developing a list of questions to ask about them. The next task, then, was to identify the qualities of cities and characteristics of typical American suburban sprawl that would give some basis for understanding preference structure in these contexts. I developed a list of factors that seemed to be important to represent and survey respondents about. The factors roughly divided into two groups, natural/aesthetic and social. The first group included elements representing complexity versus monotony, natural versus human-made components, traditional versus nontraditional building types, and sensitivity to visual clutter. The second group represented socially based factors like conduciveness to family life, human density (people occupying public spaces) versus building density (compact buildings), and privacy versus public and communal space.

Pretest

Coming up with a set of images that seemed to capture a range of characteristics about urban versus sprawl environments required a bit of trial and error. The first step was to conduct a "pretest" survey of seven selected images—three representing various characteristics of sprawl and four representing various dimensions of urban environments. The images were selected to represent the two sets of factors identified above, and included such attributes as varying levels of cars, people, pavement, and natural elements; visual clutter; high, medium, and low densities; homogeneity versus heterogeneity with respect to use and building form; and pedestrian oriented versus automobile dependent environments.

Fifteen respondents were interviewed by a graduate research assistant and asked to give their general impressions of the images. The results of these interviews were very helpful in identifying what elements or variables were being perceived by respondents. This was useful for the subsequent development of semantic differential scales of various qualities and characteristics to be used in the final survey. The results of the pretest were also helpful for making a final selection of appropriate images. Some of the images appeared to be too distracting because they showed urban scenes that were too extreme along one dimension, for example, showing places that were intensely crowded. Some images were problematic because there was one particular feature that was standing out too starkly. Other images were successful in that they seemed to be drawing out discussion in precisely the terms anticipated.

Final Selection

The number of images that could be included in the survey was eight, based on the limitations involved in doing a nationwide survey in Web television format. This number was arrived at by considering the number of questions to be asked for each image. It was felt that five questions, presented as semantic differential scales, needed to be asked about each image in order to capture the complexity of preference. Thus, 40 questions on image evaluation was considered to be maximum.

The final set of images is presented in figures 8.1 through 8.8.[15] The first three are images that represent different aspects of sprawl: Walmart

Figure 8.1 Walmart Store
Source: Author

(figure 8.1), characterizing automobile dependence, large expanses of asphalt, and lack of greenspace; a single-family housing subdivision (figure 8.2), representing single-use development and environmental degradation; and signage on an arterial (figure 8.3), representing visual clutter and lack of "sense of place." Note that the characteristics these images of sprawl are meant to convey are only hypotheses—obviously the degree to which people approve or disapprove of these environments along these lines will vary.

Images 4 through 7 capture several dimensions of urbanism: An American downtown (Houston, figure 8.4), showing the vast expanses of "lost space" in the form of parking lots contrasted with modern high rise office buildings; an American inner city location (Washington DC, figure 8.5) with some greenery (street trees) and traditional American Victorian facades; a compact European urban environment (Paris, figure 8.6) with very little green and no people; and a compact European urban environment (Brugge, Belgium, figure 8.7) with no green and heavily populated. The final image (figure 8.8) represents

Figure 8.2 Suburban Development

Source: Author

an in-between category, an urban environment at a mid-density level with significant amounts of green. The image is of Margarethenhohe, Germany, one of the earliest garden cities in Europe.

Two semantic differential scales were used:

Scale 1
Beautiful/Ugly
Exciting/Boring
Calming/Distressing
Safe/Unsafe
Would like to live near here/Would not like to live near here

Scale 2
Beautiful/Ugly
Calming/Distressing
Safe/Unsafe
Would like to live here/Would not like to live here
A good place to raise a family/Not a good place to raise a family

Figure 8.3 Signage Along Arterial

Source: Author

Figure 8.4 American City, Downtown

Source: Author (Original survey image was of Houston, TX.)

Each dimension was asked along a 5-point scale, from strongly agree to strongly disagree, and with "neutral/do not know" as a mid-scale option.

There are some subtle changes between the two scales. For the first scale it was relevant to ask about living *near* rather than living *at* the location. This made sense in places that clearly did not include residences. The second scale, in contrast, was used for images where the possibility existed that residential uses were included. For these images, the respondent was asked about living at (rather than near) the location. In addition, because of its residential possibilities, the respondent was asked about whether this would be a good place to raise a family. The addition of this question required the elimination of the "exciting—boring" question from scale 1, a question considered to be the least important for the more residential context. Scale 1 was applied to images 1 and 3 from the "sprawl" group, and image 4 from

Figure 8.5 American City

Source: Author (Original survey image was of Washington, DC.)

the "urban" group. Scale 2 was applied to image 2 from the "sprawl" group, and images 5–7 from the "urban" group, as well as image 8 that represented the mid-density option.

Explanatory Variables

In addition to understanding the various dimensions of preference for living environments, I wanted to examine what the underlying variables associated with preferences might consist of. As reviewed above, a number of variables have been correlated with preference, including race, income, gender, and stage in the life cycle, although not with a large degree of consistency. The explanatory variables used in this study were organized into four groups—sociopsychological (attachment and locus of control); attitudes toward environmental and residential issues; background characteristics; and location/region of respondent. As a general rule, in the absence of previous compelling survey data that

Figure 8.6 European City, High Density

Source: Author (Original survey image was of Paris.)

would indicate a particular hypothesized relationship, it was hypothesized that respondents would be more likely to favor what they are most familiar with, and disfavor what they are most unfamiliar with.

Social/Psychological

Two aspects of the sociopsychological dimension of respondents were hypothesized as potentially contributing to preferences about cities and sprawl. First, a set of four questions were used as a measure of local neighborhood attachment. Second, a set of six questions were used to measure "locus of control." Both measure personal dimensions of living experience.

Attachment is used commonly in sociological survey work. Attachment is a form of community sentiment, and can be separated from the more functionally based notion of community evaluation. It can be defined as the degree to which residents feel they are a part

Figure 8.7 European City, High Density

Source: Author (Original survey image was of Brugge, Belgium.)

Figure 8.8 European City

Source: Author (Original survey image was of Margarethenhohe, Germany.)

of their local community or neighborhood. Based on related multi-dimensional measures of neighboring and sense of community developed by Skjaeveland, Garling, and Maeland and McMillan and Chavis respectively, four questions were used in the survey to gauge level of attachment, which are listed in table 8.2. The results of another survey that looked at the relationship between attachment and acceptance of traditional urbanism found no relationship. This study expands upon this to include the association between attachment and preference for different types of living environments. It was hypothesized that the more attached the respondent was to their local environment, the more accepting they would be of sprawl conditions. The reasoning is that attachment is a mechanism used for adapting—higher levels of attachment have the ability to soften adverse conditions.[16]

Locus of control is a different type of sociopsychological variable. It is a personality construct that refers to how a person positions the "location" of events. The locus of events can either be internal, in which case it is determined by the behavior of the individual, or external, in which case it is determined by luck, fate, or other external set of circumstances. The locus of control idea was transformed into a personality test by Julian Rotter. The six questions used here are adapted from it, and are used to determine whether a respondent believes they control their own destiny, or whether they believe actions are determined by external forces that they cannot control. It was hypothesized that there would be some association between external locus of control and a feeling that sprawl conditions are particularly disliked. In other words, if people think the basic conditions of sprawl are ugly, they might also think that the ability to change these conditions are beyond their control. If, on the other hand, a dislike of the common conditions of American living environments is associated with internal locus of control, this would point to cognitive dissonance—an inconsistency between attitude (that it is possible to control one's own environment) and behavior (the failure to change something that is disliked).[17]

Environmental Attitudes

A second set of explanatory variables consisted of measures of the attitudes people have toward cities and environmental issues more generally. The seven questions are listed in table 8.2. Note that several of these questions are identical to the questions used in a previous survey of residential preferences. They relate to the everyday living experience of American cities and suburbs—how people feel about driving, sprawl, and cities more generally. Two questions were related to lot size and the willingness of respondents to live more compactly.[18]

The associated hypotheses were that if respondents are willing to live on a smaller lot in exchange for access to goods and services, or if, conversely, they are not willing to give up a big lot even if it means more driving, then these attitudes would correlate with preferences about compact urban living. Two additional questions were related to transportation alternatives. It was hypothesized that respondents sensitive to spending too much time in their cars, or, relatedly, those who approved of government expenditure on public transit, would also be more likely to view compact urban living (i.e., cities) in positive terms and sprawl conditions in negative terms.

Three additional questions about environment were used to gauge opinions about living environments in very general terms—the respondent's opinion of sprawl, of American cities, and of the effect of sprawl on the environment. In all cases, a straightforward hypothesis was that the more the respondent agreed with the notion of sprawl being a problem or of sprawl being harmful to the environment, the more the respondent would be likely to accept urban living, or, on the other hand, to dislike the conditions of sprawl.

Background Characteristics

A wide range of background characteristics were available for each respondent, including the standard socioeconomic characteristics of age, gender, race, ethnicity, income level, educational background, employment status, marital status, presence of children, and housing tenure. A selected number of socioeconomic characteristics for the survey respondents is summarized in table 8.3.

As noted in the literature review above, some resident background characteristics have been correlated with particular preferences; however, there is no consistent, overriding association that stands out in the survey research that would be applicable to the preferences being evaluated here. Previous, similar survey work showed no association between living preference and gender, age, presence of children, work status, and virtually all other background characteristics (although that study lacked population diversity, making it difficult to assess the differences among population subgroups).[19]

Unfortunately, the population surveyed in this study was not diverse in terms of race and ethnicity in order to analyze that aspect of preference variation.[20] It was possible to hypothesize relationships with respect to other variables, however. First, there are some indications that women would be more likely than men to favor urban environments, given their often greater requirement for access to daily life needs and the tendency for men to view residential environments as places of "refuge." On the other hand, the presence of children in the home is likely to be associated with residents who are less inclined to view urban environments favorably, due to the association between urban living and lack of safety. Related to this, it is reasonable to predict that younger respondents would be more conducive to urban environments because they view them as exciting, while aspects of sprawl might be negatively viewed as particularly mundane. Older residents may find urban environments too threatening, and find sprawl characteristics relatively less threatening.[21]

With regard to educational background, income level, and employment status, the preference for urban living would most likely follow the conventional correlation between socioeconomic status and suburban living. In the American metropolitan region, since higher income groups in most areas tend to be suburban, it would seem likely that income, tolerance for sprawl, and dislike of urban living would all be correlated. By association, it is reasonable to hypothesize that this same relationship would hold for educational level and employment status.

A final set of background characteristics looked at the previous living conditions of respondents. Two different variables were used. First, respondents were asked to indicate the type of neighborhood they lived in as a child, and second, the type of house they lived in as a child. Neighborhood types consisted of new suburb (built after World War II), older suburb (built before World War II), small town, and "in the city." Housing types consisted of single-family house on a big lot, single-family house on a small lot, duplex, townhouse or row house, or apartment building. It was hypothesized that respondents who had a background of more urban, compact living would be more favorable toward urbanism. Further, those with a "small town" background would be most likely to be unfavorable toward urbanism.

Location/Region

A final set of explanatory variables consisted of the locational attributes of the respondent. There were three aspects: the current housing type of the respondent, the location of the respondent with respect to metropolitan location, and the broader regional location. The statistics for these categories are shown in table 8.4.

The housing type of the respondent gave some indication about the character of the neighborhood of the respondent. Metropolitan location consisted of five different categories: central city location for a large city, central city location for a smaller city, suburban location for a large city, suburban location for a smaller city, and nonmetropolitan location. There were nine different regional locations: New England, Mid-Atlantic, East-North Central, West-North Central, South Atlantic, East-South Central, West-South Central, Mountain, and Pacific, which could be condensed to four: Northeast, Midwest, South, and West. Again, the hypotheses were based on the rationale that respondents will be more favorable toward what they are most familiar with in their current environment, and less favorable toward environments they are unfamiliar with. Thus, urban favorability would likely be associated with those who live in more urban parts of the country, notably the central city locations of large metropolitan areas.

Method

The survey was conducted by a marketing research company called Knowledge Networks.[22] To use the company's description, it provides "a revolutionary new single source marketing information system," conducting surveys for government, academic, and nonprofit organizations interested in research, as well as surveys for product marketing.

Knowledge Networks is fairly unique in that it has created and maintains a large national panel of households and equipped it with interactive television devices. Using these devices, respondents from around the country share their opinions in regular multimedia surveys. In this way, the company combines traditional probability sampling methodology with the multimedia capabilities of the Internet. This is a significant advantage over conventional Web surveys, which rely on users who already have Internet access. By providing free hardware and Internet access to all selected households, Knowledge Networks offers better population coverage.

What Knowledge Networks sells, then, is a panel. The panel, which is updated quarterly, is recruited using Random Digit Dialing (RDD) techniques. The sample frame consists of the entire U.S. telephone population, excluding numbers not in the WebTV Internet Service Provider network, which is about 6–8 percent of the U.S. population. After numbers are randomly selected, a mailing and telephone recruitment process is used. Recruitment interviews are conducted, where potential respondents are informed that, in exchange for responding to short surveys at the rate of about once per week, they will be given a WebTV set-top box and free monthly Internet access. There is a remote keyboard and point device. Downloads of surveys do not require any user intervention.

Extensive background data on panelists, including all members of the household aged 18 and over, is collected. Knowledge Networks claims to offer a highly representative sample of the U.S. population. The company tracks U.S. population on age, race, Hispanic ethnicity, geographical region, employment status, and other demographic characteristics. However, the company is well aware that, despite efforts to correct for known sources of deviation from equal-probability design, there are sources of survey error that are an inherent part of the process (i.e., RDD sampling rates are proportional to the number of phone lines in the household). To correct this, a sophisticated system of poststratification weights is applied.

For the survey used in this study, a random sample from the Knowledge Networks national panel was selected. Five sampling strata were used:

1. Central City MSA, population of one million or more;
2. Suburban MSA, population of one million or more;
3. Central City MSA, population up to 999,999;
4. Suburban MSA, population up to 999,999;
5. Nonmetropolitan (rural).

In each of these strata, an equal number of cases were sampled in each of the nine Census regions. Selection probabilities by strata were calculated and applied to the base panel sampling weight to account for differential sampling by strata. The final expected sample size for the survey was 1,250, or 250 completed surveys for each of the five strata. Assuming a 65 percent response rate, approximately 1,925 panel members were selected and assigned to the survey. Following standard Knowledge Networks procedure, the panelists received notification on their Internet Appliance (e-mail) that a survey was available for them to complete.

The survey was fielded between November 8, 2001 and December 3, 2001 in two waves. A reminder e-mail was sent to all survey invitees who had not completed the survey within three days after the initial invitation. There were 941 completed surveys in the first wave, and 503 in the second. The completion rate in both waves was 80 percent. There was some variation in completion rates among the different strata, ranging from 77 percent for large central city MSA's, and 85 percent for nonmetropolitan (rural) areas. In total, a random sample of 1,444 Knowledge Networks Panel Members completed the survey.

Findings

Table 8.1 lists the responses along two different scales for the eight images of sprawl and urbanism. The 5-point scale was condensed into three categories in order to present a clearer overview of significant differences.

Table 8.1 shows that a substantial majority of respondents found the commercial aspects of sprawl (figures 8.1 and 8.3) to be "ugly" and they agreed substantially that they would not want to live near these places. Approximately one-third of the respondents found the third dimension of sprawl—single-use suburban housing—to be ugly. Out of the

Table 8.1 Images Representing Dimensions of Living Preference—Dependent Variables

Scale 1	*Figure 8.1 (Walmart)*	*Figure 8.3 (Signage)*	*Figure 8.4 (Houston)*
Beautiful	8.65	2.42	8.59
Neutral/don't know	33.4	18.42	27.0
Ugly	56.9	78.19	63.99
Exciting	12.05	12.05	29.50
Neutral/don't know	38.3	37.12	34.6
Boring	48.82	49.65	35.04
Calming	11.57	3.60	3.60
Neutral/don't know	44.7	26.25	23.6
distressing	42.66	69.39	72.10
Safe	20.50	9.49	4.57
Neutral/don't know	50.1	39.63	34.1
Unsafe	28.67	49.79	60.74
Would like to live near here	14.20	7.34	5.75
Neutral/don't know	23.6	15.93	15.37
Would not like to live near here	61.49	75.97	78.46
Cronbach's Alpha reliability score	.8551	.8680	.8027

Continued

Table 8.1 Continued

Scale 2	*Figure 8.2 (Suburban housing)*	*Figure 8.5 (Washington, D.C.)*	*Figure 8.6 (Paris)*	*Figure 8.7 (Brugge)*	*Figure 8.8 (Margarethenhohe)*
Beautiful	42.80	38.64	35.38	31.65	73.96
Neutral/don't know	28.88	27.29	30.47	31.44	18.63
Ugly	27.70	33.17	33.17	36.15	6.51
Calming	52.36	20.98	23.75	10.32	67.10
Neutral/don't know	30.40	36.22	41.00	35.11	24.45
Distressing	16.41	41.97	34.42	53.60	7.82
Safe	51.87	19.39	22.09	13.02	56.16
Neutral/don't know	40.51	48.89	51.18	46.40	36.98
Unsafe	6.78	31.16	25.63	39.47	6.16
Would like to live here	41.14	18.35	18.01	12.67	52.07
Neutral/don't know	24.93	20.91	20.57	19.60	22.71
Would not like to live here	33.24	60.18	60.53	67.10	24.31
A good place to raise a family	57.21	15.58	11.78	5.88	54.16
Neutral/don't know	30.26	34.35	35.46	31.58	32.83
Not a good place to raise a family	11.98	49.38	51.94	61.35	12.26
Cronbach's Alpha reliability score	0.9090	0.8955	0.8869	0.8675	0.8967

three, signage was by far the least liked dimension in terms of ugliness, distress, safety, and living proximity. Walmart was almost identically ranked with signage along the boring-exciting dimension.

It is interesting that downtown Houston was rated similarly to the first two dimensions of sprawl. Respondents ranked the image the most distressing environment out of all eight images. It was not considered to be as ugly as the arterial signage image (figure 8.3), but it was considered to be significantly less safe and ranked similarly in terms of willingness to live close by.

Preferences along the dimensions of urbanism revealed some interesting variability. Of all urban images, downtown Houston (figure 8.4) was by far the most disliked, while Margarethenhohe (figure 8.8) was by far the most favorably rated. Margarethenhohe was considered to be almost twice as "beautiful" as the images of Washington, DC, Paris, and Brugge, and significantly more beautiful than the suburban housing development. It ranked very similarly to the suburban housing development in terms of safety and as a place to raise a family, but was considered significantly more calming and rated higher as a place where respondents would like to live. Margarethenhohe was the only image to be ranked by fewer than 10 percent of the respondents as ugly, distressing or unsafe. Houston, by contrast, was ranked by more than 60 percent of respondents along these same three dimensions.

The images of Washington, DC, Paris, and Brugge represented similar levels of urbanism, and were ranked similarly along the beautiful-ugly scale. The image of Brugge, which differed from the others especially in terms of the presence of people, was considered to be the most distressing image of the three, although not as distressing as downtown Houston. Brugge was also considered significantly less safe, and less liked as a place to raise a family. Further, a relatively small percentage of respondents considered any of these three levels of urbanism to be places they would like to live, and an even smaller percentage thought these places would be a good place to raise a family. Margarethenhohe, the German garden city, was substantially different than the urban examples in this regard.

It is also interesting that the respondents were evenly divided on the issue of whether the urban places of Washington, DC, Paris, and Brugge were beautiful or ugly. Roughly one-third ranked these three images as beautiful, one-third ranked them as ugly, and one-third ranked them as "neutral/don't know." Many respondents were indifferent or neutral to these images of urbanism along other dimensions as well. More than a third were neutral in terms of calmness and family raising. For

Paris, a majority of respondents were neutral on the issue of safety (51 percent) and 41 percent were neutral on the issue of calmness. Overall, substantial numbers of respondents were indifferent to these three images of urbanism along the beautiful-ugly, calming-distressing, and safe-unsafe scales, but respondents were much clearer about whether they would like to live there and whether these were considered good places to raise a family.

Respondents were neutral on a number of scales for many of the images. Close to one-third of the respondents were neutral toward Walmart in terms of beauty and excitement, and an even greater number were neutral in terms of calmness and safety. A third of the respondents were neutral toward Houston in terms of excitement and safety. For suburban housing, 40 percent were neutral in terms of safety, and close to a third were neutral in terms of beauty, calmness, and family raising.

Tables 8.2, 8.3, and 8.4 list the variables and responses for the various categories of explanatory variables. A few of the more significant differences can be highlighted. As expected, respondents were attached to their neighborhoods, although they were fairly evenly divided on the question of whether or not they would like to move to another location. Close to 80 percent disagreed with the statement that they did not feel "at home" in their neighborhood.

Many respondents seemed ambiguous on the locus of control questions. A substantial majority were neutral or ambiguous (partly agree, partly disagree) on the question of whether heredity determines personality, as well as whether being at the right place at the right time is essential for "getting what you want in life." However, a sizable majority disagreed that people are lonely because they are not "given the chance to meet new people," and that school success is a matter of socioeconomic background, both indicators of internal locus of control. Questions about chance and the idea that "something" always gets in the way of plans were fairly evenly distributed, although most respondents disagreed that chance was not relevant.

There was more variability across the locus of control scale than the attachment scale, and this is reflected when an index of scaled responses was constructed. The Cronbach alpha reliability coefficient for the four attachment measures was 0.8126, while the coefficient for the six locus of control measures was only 0.3746. This prohibited the use of an index for a combined locus of control measure. A combined index for the attachment scale was used, however.

Environmental attitudes shown in table 8.2 reveal a respondent population mostly unwilling to live on a smaller lot in exchange for

Table 8.2 Sociopsychological Characteristics and Environmental Attitudes—Explanatory Variables

	Response (Percent)[a]		
	Disagree	*Partially Agree/Partially Disagree or Neutral*	*Agree*
Sociopsychological Factors Attachment[b]			
I plan to remain a resident of my neighborhood for a number of years	13.92	17.04	68.90
I feel like I am strongly attached to my neighborhood	24.38	28.05	47.51
Given the opportunity, I would like to move to another location	39.06	27.63	33.18
I don't feel at home in my neighborhood	78.88	13.37	7.69
Locus of Control[c]			
Heredity determines most of a person's personality	27.29	50.69	21.82
Chance has little to do with being successful	37.68	37.05	25.28
Whatever plans you make, there is always something that will cross them	39.68	37.19	22.92
Being at the right place, at the right time is essential for getting what you want in life	25.62	43.21	30.92
School success is mostly a result of one's socioeconomic background	56.37	28.25	15.30
People are lonely because they are not given the chance to meet new people	71.95	20.64	7.21
Environmental Attitudes			
I would be willing to live on a smaller lot in exchange for more parks, shops, or schools that I could walk to	62.33	20.91	16.48
I prefer to live in a suburb on a big lot, even if it means more driving	24.79	24.52	50.49
I spend too much time in my car	60.18	25.00	14.55
The government should invest in public transportation instead of building more roads and highways	33.10	38.71	27.98
Suburban sprawl is not a big problem where I live	29.98	26.45	43.28
Suburban sprawl is harmful to the environment	18.97	42.87	37.82
Generally speaking, American cities are not very good places to live	38.99	33.73	26.94

[a] Missing responses not included in table.
[b] Cronbach's Alpha reliability coefficient = 0.8126.
[c] Cronbach's Alpha reliability coefficient = 0.3746.

more parks, shops, and schools to walk to, mostly agreeing that a big lot is preferred, even if it means more driving, and disagreeing that they spend too much time in their car. Respondents were more evenly divided on the general question of whether American cities are good places to live, as well as on the question of whether government should invest in public transportation. A substantial number of respondents were ambivalent about whether sprawl was harmful to the environment (about 43 percent), although an equal percentage agreed that sprawl was a "big problem" in their location.

The resident characteristics listed in table 8.3 show a population that is more homogeneous than the American population overall, especially in terms of race and ethnicity. The sample contained only about 4 percent Black/African American compared to 12 percent nationally, and only 2 percent Hispanic compared to 13 percent nationally. The sample is reasonably representative of the U.S. population in terms of gender (although it is slightly more male), household income, and education. In terms of age, the sample is older than the general American population (about 18 percent of the U.S. population is aged 45–59, for example, compared with 32 percent in the sample). The sample is also slightly higher in terms of its married population (64 percent in the sample but 54 percent nationally). Finally, the sample contains a lower percentage of households who rent—18 percent in the sample compared to 34 percent for the U.S. population as a whole.

Table 8.4 presents the attributes of the sample in terms of locational characteristics. Most respondents said they lived, as a child, in a small town, and that they lived in either a single-family home on a big lot or a single-family home on a small lot. Respondents were fairly well distributed on the two main locational variables, both of which were used to stratify the sample. Regionally, the sample was slightly skewed toward the South (representing 33 percent of the sample).

Table 8.5 shows a crosstabulation between current housing type (from table 8.4) and the two background characteristics presented in table 8.3—neighborhood lived in as a child and housing type lived in as a child. Of the 86 percent of the sample that currently live in a single-family detached dwelling, a significant majority said they lived in a single-family home on a big lot in a small town. Further, crosstabulation between the two background characteristics of "neighborhood lived in as a child" and "housing type lived in as a child" (not shown in the table) revealed that the largest category consisted of residents who had mostly lived in a small town in a single-family detached house on a

Table 8.3 Resident Characteristics—Selected Explanatory Variables

	Frequency	*Percent*
Gender		
Male	768	53.19
Female	676	46.81
Age		
18–29	163	11.30
30–44	455	31.53
45–59	457	31.67
60+	368	25.50
Race		
White	1257	89.98
Black/African American	53	3.79
American Indian or Alaska Native	24	1.72
Asian/Pacific Islander	33	2.36
Other	29	2.08
Ethnicity		
White, Nonhispanic	1245	88.68
Black, Nonhispanic	52	3.70
Other, Nonhispanic	81	5.77
Hispanic	26	1.85
Marital status		
Married	920	64.43
Single, never married	233	16.32
Divorced	173	12.11
Widowed	85	5.95
Separated	17	1.19
Education		
Less than high school	89	6.18
High school	350	24.31
Some college	619	42.99
Bachelor or higher	382	26.53
Children in Household		
Children (17 or younger) living in household	381	26.39
No children present	1063	73.61
Household Income		
Under 30,000	309	25.69
30,001–50,000	328	27.27
50,001–75,000	325	27.02
Over 75,000	241	20.04
Housing Tenure		
Rent	259	17.94
Own	1109	76.80

Continued

Table 8.3 Continued

	Frequency	*Percent*
Neighborhood Type Lived in as Child		
New suburb	254	17.59
Older suburb	127	8.80
In the city	252	17.45
Small town	613	42.45
Other	195	13.50
Housing Type Lived in as Child		
Single-family/big lot	673	46.61
Single-family/small lot	472	32.69
Duplex/two-family house	88	6.09
Townhouse/row house	31	2.15
Highrise apartment	30	2.08
Other	147	10.18

Table 8.4 Locational Characteristics—Selected Explanatory Variables

	Frequency	*Percent*
Housing Type (Current Residence)		
Single-family detached	1018	85.62
Single-family attached	57	4.79
Apartment	15	1.26
Condominium or co-op	6	0.50
College dormitory	1	0.08
Manufactured or mobile home	89	7.49
Other	3	0.25
Region		
Northeast	324	22.44
Midwest	321	22.23
South	470	32.55
West	329	22.78
Metropolitan location		
Central City, 1,000,000 or more	270	18.70
Central City, under 1,000,000	305	21.12
Suburban, 1,000,000 or more in metro area	325	22.51
Suburban, under 1,000,000 in metro area	237	16.41
Nonmetro (rural)	307	21.26

Table 8.5 Background Characteristics Related to Current Housing Type

	Current Housing type—Single-Family Detached	
	Frequency	*Percent*
Neighborhood Lived in as Child		
New suburb	196	19.25
Older suburb	85	8.35
In the city	162	16.11
Small town	432	42.44
All other	140	13.75
		100.00
Housing Type Lived in as Child		
Single-family/big lot	509	50.00
Single-family/small lot	348	34.18
Duplex, townhouse, rowhouse, high rise	68	6.68
Other	91	8.94
		100.00

big lot. There were 341 respondents, or approximately one-third, who fit this category.

Explanatory Analysis

There were four groups of explanatory variables used in this survey—sociopsychological attitudes (attachment and locus of control); attitudes toward environmental and residential preference; background characteristics; and location/region of respondent. Bivariate and multivariate analyses were used to try to reveal what variables might have a significant explanatory effect on living preferences.

To begin with, responses that measured neighborhood attachment and locus of control were analyzed by computing crosstabulation scores for all sociopsychological questions separately, as well as for the attachment index (the locus of control measure was not indexed given low average inter-item covariance). These results are not reported here because they were *all* insignificant. Respondents were uniformly attached to their neighborhoods (table 8.2), but this seemed to have no relationship to the opinions of the respondents on urbanism and sprawl.

Respondents had varying ideas about external versus internal locus of control (table 8.2), but none of these variations could be used to explain the variation in opinion on the eight dependent variables.

There was more association discernible using the "environmental attitudes" variables. For each of the eight images, the five semantic scale variables were combined into a single scaled variable. These index scores were then divided into five quantiles ranging from low to high, and these were then used in the crosstabulations. The strongest variations, shown for two different variables, are given in table 8.6. Significant relationships were for images of urbanism—there were no associations found between environmental issues and the images of sprawl.

Table 8.6 shows two different directions of association. For the question about willingness to live on a smaller lot in exchange for access to facilities, respondents who were less favorable toward any of the four listed dependent variables were also less willing to trade lot size for access to facilities.[23] In other words, if respondents were more favorable toward any of these urban environments, they were also more willing to trade lot size for facilities. What is surprising is that this relationship held whether or not the dependent variable was an image of Houston, a low ranked image, or the image of Margarethenhohe, which was the highest ranked image.

Correlation for the second explanatory variable shown in table 8.6 is consistent with the first. For a slightly different set of dependent variable images, the association with the second variable revealed that the more respondents agreed that American cities are not very good places to live, the less favorable their opinion of urban environments was. The more respondents disagreed that American cities are not good places to live, the more they were willing to accept a smaller lot size in exchange for access to facilities.

These findings are as expected. What is surprising is that the relationships were the same for widely different representations of city life. Upon further investigation, crosstabulations between different dependent variables revealed strong associations between even seemingly different aspects of urbanism. For example, there was strong correlation between Houston (image 4) and Paris (image 6) for the semantic scale of "would like to live here—would not like to live here" (probability was 0.000, Pearson chi2 statistic was 338.85 and gamma was 0.3773). Similar strong associations were found on other semantic scale questions between Houston and Paris, and between Houston and Brugge as well.

Table 8.6 Summary of Crosstabulation Statistics[a]

Dependent Variable[b]	*Explanatory Variable*	*Chi Square*	*Cramer's V*	*Gamma*	*Kendall's tau-b*
Figure 8.4 (Houston)	I would be willing to live on a smaller lot in exchange for more parks, shops or schools that I could walk to.	69.89	0.1270	-0.2080	-0.1559
Figure 8.5 (Chicago)	"	79.11	0.1351	-0.2359	-0.1760
Figure 8.7 (Brugge)	"	69.01	0.1262	-0.2012	-0.1500
Figure 8.8 (Margarethenhohe)	"	106.83	0.1570	-0.2234	-0.1657
Figure 8.4 (Houston)	Generally speaking, American cities are not very good places to live.	134.76	0.1764	0.2855	0.2141
Figure 8.5 (Chicago)	"	79.57	0.1355	0.2168	0.1619
Figure 8.6 (Paris)	"	93.69	0.1471	0.2290	0.1711
Figure 8.7 (Brugge)	"	98.99	0.1512	0.2467	0.1841

[a] All crosstabs shown had probabilities less than 0.001.

[b] Questions for dependent variables along 5 scales were combined into a single-scaled variable.

Background characteristics were next looked at as possible explanatory indicators of differences in living preferences. Race and ethnicity could not be evaluated because of the low representation in the sample of nonwhite and Hispanic groups. Income, marital status, presence of children, housing tenure, and gender showed no explanatory effect in bivariate analyses for all dependent variables.

The resident characteristics that did have an explanatory effect were age and education level. Specifically, education level was inversely associated with preference for the three urban images shown in figures 8.5 (Washington, DC), 8.6 (Paris) and 8.7 (Brugge). The higher the education level, the more favorable respondents were toward these images. There were no differences with respect to images of sprawl.

Rankings for several different related images were combined into two indexes to summarize the differences in preferences by age category. The results are listed in table 8.7. Figures 8.1–8.4 were combined for the "sprawl" index and figures 8.6 and 8.7 for the "Europe" index. These indices, which had high alpha reliability coefficients, were sorted and divided into quantiles for analysis. What the table indicates is that younger respondents had a much more positive rating of both sprawl and the two images of European urbanism, while older respondents

Table 8.7 Crosstabulations, Age, and Preference for (1) Sprawl and (2) European Urbanism[a]

Quantiles[b] of (1) sprawl (2) Europe	*Age (Frequency/Percent)*				
	18–29	*30–44*	*45–59*	*60+*	*Total*
1	(1) 81/50	(1) 131/29	(1) 106/23	(1) 81/22	(1) 399/28
(Positive)	(2) 70/43	(2) 127/28	(2) 102/22	(2) 65/18	(2) 364/25
2	(1) 40/24	(1) 118/26	(1) 113/25	(1) 88/24	(1) 359/25
	(2) 47/29	(2) 124/27	(2) 110/24	(2) 85/23	(2) 366/25
3	(1) 32/20	(1) 117/26	(1) 129/28	(1) 81/22	(1) 359/25
	(2) 26/16	(2) 113/25	(2) 120/26	(2) 108/29	(2) 367/25
4	(1) 10/6	(1) 89/20	(1) 109/24	(1) 118/32	(1) 326/23
(Negative)	(2) 20/12	(2) 91/20	(2) 125/27	(2) 110/30	(2) 346/24
Total	163/100	455/100	457/100	368/100	1443/100

[a] All crosstabs shown had probabilities less than 0.001. Statistics for (1) are Pearson chi2 = 78.2628, Cramer's V = 0.1345, gamma = 0.2158, and Kendall's tau-b = 0.1593. Statistics for (2) are Pearson chi2 = 61.9483, Cramer's V = 0.1196, gamma = 0.2172, and Kendall's tau-b = 0.1604.

[b] Rankings for dependent variables along 5 scales are combined, figures 8.1–8.4 for the "sprawl" index and figures 8.6 and 8.7 for the "Europe" index. Alpha reliability coefficient for the sprawl index = 0.8744, and for the Europe index = 0.8999. These combined scales were each sorted into quantiles. Quantiles with the same scaled scores were not split, which accounts for the differences in quantile sample sizes.

had a more negative rating for both. In terms of percentages, the younger category was more than twice as favorable (for either index) as the older category. Further, a third of the 60+ age categories ranked sprawl and urbanism low, but negative ranking among the 18–29 age group comprised only 6 percent and 12 percent of the sample for sprawl and urbanism rankings respectively.

Another set of explanatory variables pertained to background living characteristics—the neighborhood and housing type the respondent lived in as a child. Table 8.5 showed how respondents mostly identified themselves as coming from a background of single-family homes in a small town. Perhaps not surprisingly, this group was also most likely to dislike all forms of urbanism. Consistent with earlier findings, there seemed to be little distinction whether that urbanism was in the form of Houston, Paris, or Brugge, or whether the development was in the form of sprawl. Those growing up in a small town in a single-family home had fewer positive feelings about development of any type, relative to those who did not grow up in a small town.

For regional location, there was some indication that respondents living in the Northeast were more negative about sprawl than respondents in the South. Specifically, table 8.8 indicates that those in the South and Midwest were more likely than those in the Northeast to have a positive view of Walmart, while those in the Northeast, compared to all other regions, were more likely to have a negative view of Walmart.

Table 8.8 Crosstabulation, Walmart, and Region[a]

Quantiles[b] of Walmart (Figure 8.1)	*Region (Frequency/Percent)*				
	Northeast	*Midwest*	*South*	*West*	*Total*
1 (Positive)	82/25	111/35	188/40	101/31	482/33
2	43/13	61/19	77/16	68/21	249/17
3	121/37	94/29	118/25	103/31	436/30
4 (Negative)	78/24	55/17	87/19	57/17	277/19
Total	324/100	321/100	470/100	329/100	1444/100

[a] Crosstabulation had a probability of less than 0.001. Statistics are Pearson chi2 = 34.5081, Cramer's V = 0.0893, gamma = -0.0857, and Kendall's tau-b = -0.0634.

[b] Rankings for dependent variable Walmart (figure 8.1) along 5 scales are combined, and combined scale index is sorted into quantiles. Quantiles with the same scaled scores were not split, which accounts for the differences in quantile sample sizes.

In terms of metropolitan location, crosstabulations with Houston and Paris, shown in table 8.9, indicate that respondents in central cities in large metropolitan areas (population of one million or more) were more likely to view both types of urbanism more favorably than those in other locations. Respondents in large central cities tended to be the most differentiated group in terms of preferences for all images. The table also shows that residents in large central cities were the most likely to view Paris and Houston similarly. By contrast, suburban residents had much greater variation in how they viewed urbanism—while 19 percent were in the upper quantile for Houston, 31 percent were in the upper quantile for Paris. This can be contrasted with respondents in the central city, where an equal percentage of respondents were in the upper quantile for Houston and Paris.

The final part of the analysis was to investigate multivariate relationships among the four categories of explanatory variables. Multiple regression analysis was performed where image preferences were used as dependent variables, and the independent variables consisted of attachment and locus of control, attitudes toward environmental and residential issues, background characteristics, and location/region of respondent. Ordinal variables for the semantic scales were used using a regression procedure in which categorical variables are expanded into indicator (also called dummy) variable sets by creating new variables. The dependent variables consisted of scaled index variables constructed from the 5 semantic scales for each image. Indexes constructed from combined scales (using more than one image) were also used as dependent variables.

The full spectrum of variables hypothesized as having some explanatory effect on living preferences was used in a step-wise elimination procedure. One regression result is shown in table 8.10. The dependent variable used in this case was a scaled index variable consisting of all semantic scaled responses for images 6 and 7—Paris and Brugge. The dependent variable was thus a measure of preference for what could be termed "European urbanism." The final set of variables found to be significant for predicting preference for European urbanism in a multivariate framework are listed. Only variables with a probability less than 0.01 are shown. The R-squared for the final regression model was 0.1973 (with an adjusted R-squared of 0.1758). While there is a high amount of unexplained variation, this is not unusual for a survey sample of this size.

Despite the broad range of explanatory variables available, the multivariate approach leaned predominantly on environmental attitudes as

Table 8.9 Crosstabulations, Metropolitan Location, and Preference for (1) Houston and (2) Paris[a]

Quantiles[b] of (a) Houston (b) Paris	*Metro location (Frequency/Percent)*					
	Central city, pop. > 1 mill.	*Central city, pop. < 1 mill.*	*Suburban, pop. > 1 mill*	*Suburban, pop. < 1 mill.*	*Nonmetropolitan*	*Total*
1	(1) 104/39	(1) 85/28	(1) 69/21	(1) 46/19	(1) 57/19	(1) 361/25
(Positive)	(2) 106/39	(2) 91/30	(2) 94/29	(2) 74/31	(2) 77/25	(2) 364/25
2	(1) 83/31	(1) 64/21	(1) 108/33	(1) 64/27	(1) 82/27	(1) 401/27
	(2) 74/27	(2) 85/28	(2) 84/26	(2) 54/23	(2) 86/28	(2) 383/27
3	(1) 46/17	(1) 82/27	(1) 82/26	(1) 63/27	(1) 84/27	(1) 357/25
	(2) 43/16	(2) 72/24	(2) 84/26	(2) 50/21	(2) 63/21	(2) 312/22
4	(1) 37/14	(1) 74/24	(1) 66/20	(1) 64/27	(1) 84/27	(1) 325/23
(Negative)	(2) 47/17	(2) 57/19	(2) 63/19	(2) 59/25	(2) 81/26	(2) 307/21
Total	270/100	305/100	325/100	237/100	307/100	1444/100

[a] Houston had a probability less than 0.001, Paris had a probability less than 0.01. Statistics for (1) are Pearson chi2 = 64.2291, Cramer's V = 0.1218, gamma = 0.1699, and Kendall's tau-b = .1317. Statistics for (2) are Pearson chi2 = 28.0372, Cramer's V = 0.0804, gamma = 0.1082, and Kendall's tau-b = 0.0834.

[b] Rankings for dependent variables along 5 scales were combined. These combined scales were then sorted into quantiles at break points. Quantiles with the same scaled scores were not split, which accounts for the slight differences in quantile sample sizes.

Table 8.10 Multiple Regression Analysis Results, European Urbanism as Dependent Variable[a]

Explanatory Category	*Statement*	*Response*	*Coefficient*[b]	*Standard error*	*t*	*Probability*
Environmental Attitudes	I would be willing to live on a smaller lot in exchange for more parks, shops of schools that I could walk to	Strongly agree	−0.516	0.136	−3.79	0.000
	I prefer to live in a suburb on a big lot, even if it means more driving	Strongly disagree	−0.326	0.104	−3.13	0.002
	"	Partially agree/ disagree or neutral	−0.231	0.077	−2.99	0.003
	The government should invest in public transportation instead of building more roads and highways	Strongly disagree	0.433	0.108	4.02	0.000
	"	Disagree	0.278	0.098	2.84	0.005
	"	Partially agree/ disagree or neutral	0.249	0.093	2.66	0.008
	Generally speaking, American cities are not very good places to live	Strongly disagree	−0.450	0.088	−5.11	0.000
	"	Disagree	−0.324	0.063	−5.11	0.000
	"	Partially agree/ disagree or neutral	−0.252	0.061	−4.10	0.000
Background characteristics	Age	45–59	0.305	0.085	3.58	0.000
	"	60+	0.466	0.088	5.31	0.000
Location	Metropolitan location	Nonmetropolitan (rural)	0.157	0.056	2.78	0.006

[a] R-squared = 0.1973 (adjusted = 0.1758). Obs = 1188 (observations with missing values excluded).

[b] Note that the higher the score, the less favorable the response toward the dependent variable.

significant predictors. Note that higher scores for the combined scale (combining preferences for Paris and Brugge) indicate a less favorable response for the 5 semantic scales—in the survey, a score of "1" was most favorable and a score of "5" was least favorable.[24] The stronger the preference for European urbanism—images of Paris and Brugge—the more likely respondents would be to "strongly agree" with the statement about exchanging lot size for access to facilities. Another strong predictor was the preference for living in a suburb on a big lot "even if it means more driving." Strongly disagreeing, disagreeing,[25] or being neutral on this statement showed an inverse relation with dislike for European urbanism. More favorable response toward European urbanism was predicted by disagreement about the importance of big suburban lots.

Two other environmental attitudes that had explanatory effect in a multivariate context were public investment in transit and agreement or disagreement with the statement that American cities are good places to live. Disagreement with government investment in public transit was associated with an unfavorable view of European urbanism. Disagreement that American cities are not good places to live (thus suggesting a favorable attitude toward American cities) was correlated with a preference for European urbanism.

Apart from these environmental attitude variables, the only other variables to emerge that had significant explanatory effect in a multivariate framework were age and nonmetropolitan or rural location. As in the bivariate analysis, older respondents were associated with a greater dislike for urbanism. In terms of location, respondents in a nonmetropolitan area were associated with a less favorable attitude toward urbanism.

Discussion

It is important not to overanalyze, or to read too much into, the above quantitative results. There were obvious limitations with the methodology driven by the fact that only a small number of images could be fielded for this broad-based survey. The images were not scientifically derived representations of factors to be extracted. Again, this is an exploratory study intended to provide a basis for further investigation. With this caveat in mind, in this section I propose some general impressions of the survey data reported above.

The first issue raised in reviewing these survey results is that so many respondents seem to be indifferent to images of urbanism. The fact that respondents were more indifferent along the beauty, calmness, and safety scales but less indifferent in terms of stating whether or not they would want to live there, may indicate that many respondents have a negative opinion of these places for unspecified reasons. It is possible that there are reasons for these negative views that were not included in this survey, and these would need to be further explored. Yet the fact that there was a strong relationship between European urbanism and Houston, images representing what many planners and architects would consider vastly different forms of urbanism, points to an underlying indistinctness.

The ambivalent attitudes about issues like sprawl, land, proximity to services, and public transportation also could be interpreted as reflective of a public that is largely unclear about the implications and trade-offs of different types of living environments. Consistent with other previous surveys identified earlier, this study revealed a population unwilling to make the trade-offs necessary to get what it appears to want. This either reflects a weak conviction or an inability or unwillingness to conceptualize the trade-offs involved in low density living. Either sprawl is not enough of a burden, or respondents have a general apathy about living environments in general. This is certainly indicated by the high number of respondents who were neutral or ambivalent about whether sprawl was harmful to the environment. Most respondents were either neutral or agreed that cities are not good places to live, indicating, in general, a sense of neutrality, ambivalence or disdain about the American settlement pattern.

These impressions from the data must be contrasted with the significant fact that most of the respondents claimed to have a childhood living experience consisting of a single-family house in a small town. This is, in many ways, the ideal environmental type for many Americans—a single-family house in a small town represents the ideal arrangement of land, individuality of ownership, and space for a family, but within the context of a "town" that is able to satisfy needs for goods, services, and other amenities in a positive way. In other words, an individualized house without the associated negatives of sprawl—a rural life style with an urban level of services. Many Americans ascribe to this ideal, but faced with a living environment closer to sprawl rather than a small town, it may be the case that many Americans have experienced a downgrading in the quality of their living environments, away from the small town ideal of their childhoods.

Another particularly significant outcome of the survey was the fact that so many respondents rated Brugge lower than Paris, when the only really significant visual difference was in terms of human population. Both images are of similar density, and both consist of classical architecture. The architecture shown in the Brugge image is considered to be of particularly high quality, making the city one of the most popular tourist destinations in Europe. Yet the majority of respondents found this image distressing, and indicated that they would not like to live there, nor did they consider it a good place to raise a family. This likely had to do with the presence of people in the Brugge image. The results thus suggest a sensitivity to crowds. But note that this attitude was similar for Houston that was considered even more distressing. Instead of people, the crowd aspect in the Houston image was indicated by parking lots full of cars. It is significant that in both cases the negative view of urbanism was more a matter of distress than lack of safety.

It is interesting to compare some of the attitudes expressed in this study with those of the previous study of preferences in a Dallas suburb that used some very similar questions.[26] The Dallas study was much smaller (n = 185) and limited to one area but some comparisons can still be made. The most significant difference between the two populations was in terms of income level. In the Dallas study, 78 percent of the respondents had an annual household income greater than $100,000, while only 20 percent of the respondents of this study had an income greater than $75,000. The current study sample was also older, less educated, and had fewer children present in the home. On the other variables, such as race, ethnicity and gender, there were significant similarities.

Table 8.11 compares the responses for the two surveys on questions that were almost identically worded. The higher income respondents

Table 8.11 Comparison of Responses between Two Similar Surveys[a]

	Dallas Survey (Percent Agree/Disagree)	*This Survey (Percent Agree/Disagree)*
I prefer to live in a suburb on a big lot, even if it means more driving[b]	75/18	51/25
I spend too much time in my car	50/35	15/60
The government should invest in public transportation instead of building more roads and highways	32/41	28/33

[a] This survey compared to questions reported in Talen, 2001.
[b] There is some variation in wording between the two surveys for this question.

from the Dallas survey were much more likely to prefer living on a big lot in spite of the requirement for more driving. Yet they were also much more likely to be car sensitive—that is, to agree with the statement that they spend too much time in their cars. Respondents on the two surveys were more alike in terms of government spending for public transit rather than highways, which both groups were more likely to agree than disagree with. It could be argued, then, that whereas environmental preference is strongly related to current living environment, attitudes about public transit are more uniformly distributed.

The bivariate analysis indicated that the more favorable respondents are to images of cities, the more they are willing to think of cities as good places to live, and the more they would be willing to exchange lot size for access. This makes intuitive sense. The more surprising aspect of these relationships was that it revealed that respondents did not seem to differentiate between different types of urbanism. The strong associations between Houston and other forms of urbanism seem to indicate that respondents think of urbanism unidimensionally. Paris and Houston, for example, might have been expected to be inversely related, but the fact that they were positively related indicates that respondents may be thinking of urban places, no matter what the form, in similar terms. This same phenomenon was present even for respondents living in different metropolitan locations. In fact, respondents residing in large central cities were the most likely to view both Houston and Paris favorably—two very different representations of urbanism. Again, this is surprising given the differences between them—Paris is pedestrian-oriented and enclosed, while Houston is automobile oriented and open.

Results from the bivariate analyses linking region and metropolitan location confirm conventional wisdom that people are likely to be more comfortable with what they are more used to. It is reasonable to assume that the Walmart form of development pictured in figure 8.1 contrasts more with historical development patterns in the Northeast as opposed to the Midwest and South, and is therefore likely to be more disliked. Perhaps respondents in the South and Midwest are more accustomed to Walmart as a development type and therefore less likely to oppose them. A check of Walmart store statistics at the time of the survey, found on the company Web site, confirmed this. There were 386 stores[27] in the Northeast, 800 stores in the Midwest, and 1,657 stores in the South. This translates to a per capita rate of 1 store per 139,000 population in the Northeast, 1 store per 81,000 in the Midwest, and 1 store per 62,000 in the South. Since the Northeast has half the rate of Walmart stores per capita than the South, the contention that there is a relationship between preference and familiarity is supported. Of course, this relationship could work both ways: the South and Midwest may have relatively more Walmarts per capita because there is a more favorable attitude toward shopping at them to begin with.

The multivariate analysis gave no surprises. Respondents indicating preferences along more urbanistic lines—willingness to live on a small lot, weak importance attached to big lots, willingness to invest in public transit and a favorable attitude toward American cities—were all attitudes that had a strong explanatory effect on preference for urbanism as shown in the images of Paris and Brugge. These associations are as expected. What is perhaps more enlightening is the recurrent significance attached to age—older respondents had a less favorable view of European urbanism in a multivariate context, just as in the bivariate analysis.

The multivariate analysis was interesting not only for what variables emerged as significant, but also for what variables were left out. Attachment and locus of control had no explanatory effect. The attitudes toward environment were predictably associated, but it is interesting that certain ones had stronger association than others. The fact that older respondents and rural respondents were good predictors of preference for European urbanism is consistent with the bivariate analysis, but the fact that they emerge in a multivariate framework in which other variables are controlled for speaks to the relevant strength of these associations.

Conclusion

This investigation of some perceived aspects of urbanism and sprawl, as an exploratory study, was not set up to definitively answer particular questions about American preferences. The results, instead, give certain impressions, raise certain questions, and give a clear indication that more exploration is warranted.

For those intent on reducing sprawl and increasing the urbanistic qualities of American living environments, the results from this survey could potentially be used to bolster the contention that American preferences are adaptable. While the living preferences of Americans are predictably about wanting to have it all—the benefits of small town living without the associated negatives of sprawl, essentially wanting suburban living with access to urban level services—this inconsistency and at times ambiguity suggests a preference structure that is volatile.

It may be possible to capitalize on this volatility and educate the public more clearly about living preferences and the trade-offs involved. This seems essential given the issue that Americans are not favorable about images of sprawl—images that characterize much of the environment in which Americans live. We should be concerned, in other words, about the idea that many Americans likely have a negative view of the places that surround them.

One phenomenon that may have some bearing on this disjuncture between preference and the living environments of Americans is what is known as "cognitive dissonance theory."[28] Dissonance occurs when there is an inconsistency between an attitude and a behavior, and individuals will try to rectify this by changing one or the other. In most cases, it is the attitude that will change rather than the behavior. It is conceivable that, over time, Americans could change their attitudes toward their living environments in order to avoid dissonance. After all, if Americans strongly dislike some aspects of the environments they are at least indirectly choosing for themselves—through their own behavior—there is dissonance. It could be rectified, theoretically, by changing attitudes to be more in line with behavior. To some extent, this is already reflected in the phenomenon verified in this survey that people tend to prefer what they are most familiar with.

On the other hand, there is a need to find ways to help Americans feel better about cities. We need to strengthen the quality of urbanism, but we also need to address the fact that many Americans lack a positive experience *living* in cities. Clearly, as evidenced from this survey, many Americans continue to have a low opinion of urbanism. It was particularly surprising that so many respondents in the survey rated places like Paris and Brugge, two of the most popular tourist destinations in the world, as places they would not want to live. Americans may be able to appreciate these places as tourist destinations, but they remain uncomfortable or perhaps just unaware of the experience of living in such places. This impression has led some commentators to argue that Americans lack understanding and education about cities.[29] If it is a proper goal to increase the satisfaction people have with their living environments, perhaps we should consider not only the need to help change people's living environments, but to address the educational support urban living requires.

Notes

This study was funded in part by the National Science Foundation, grant no. NSF BCS 99–05349.

1. Morton White and Lucia White, *The Intellectual Versus the City: From Thomas Jefferson to Frank Lloyd Wright* (Cambridge, MA: Harvard University Press, 1962), 17, 218; see also James Howard Kunstler, *Home from Nowhere* (New York: Simon & Schuster, 1996); Kenneth J. Jackson, *Crabgrass Frontier: The Surbubanization of the United States* (New York: Oxford University Press, 1985). Robert Fishman, *Bourgeois Utopias: The Rise and Fall of Suburbia* (New York: Basic Books, 1987).
2. David Schuyler, *The New Urban Landscape: The Redefinition of City Form in Nineteenth-Century America* (Baltimore: Johns Hopkins University Press, 1986), 24. Tony Hiss, *The Experience of Place* (New York: Alfred A. Knopf, 1991). Lewis, Mumford, "The Natural History of Urbanization," in William L. Thomas (ed.), *Man's Role in Changing the Face of the Earth* (Chicago, IL: University of Chicago Press, 1956).
3. Robert Fishman, *The American Planning Tradition* (Washington, DC: Woodrow Wilson Center Press, 2000). Morton White and Lucia White, *The Intellectual versus the City: From Thomas Jefferson to Frank Lloyd Wright* (Cambridge, MA: Harard University Press, 1962).
4. Robert W. Marans and William Rodgers, "Evaluating Resident Satisfaction in Established and New Communities," in Robert W. Burchell (ed.), *Frontiers of Planned Unit Development* (New Brunswick, NJ: Center for Urban Policy Research, 1973), 197–227. Anne B. Shlay, "Taking Apart the American Dream: The Influence of Income and Family Composition on Residential Evaluations," *Urban Studies* 23 (1986): 253–270. Reid Ewing, "Is Los Angeles-Style Sprawl Desirable?" *Journal of the American Planning Association* 63 (1) (1997): 107–126. Mark Baldassare and Georjeanna Wilson, "More Trouble in Paradise? Urbanization and the Decline in Suburban Quality-of-Life Ratings," *Urban Affairs Review* 30 (5) (1995): 690–708. Sidney Brower, *Good Neighborhoods: A Study of In-Town and Suburban Residential Environments* (Westport, CT: Praeger, 1996). Andres Duany and Elizabeth Plater-Zyberk, "The Second Coming of the American Small Town," *Wilson Quarterly* 16 (1) (1992): 3–51. Anton Nelessen, *Visions for a New American Dream* (Chicago, IL: American Planning Association. American Lives, 1995; New Urbanism Report. San Francisco, CA).
5. Ivonne Audirac, "Stated Preference for Pedestrian Proximity: An Assessment of New Urbanist Sense of Community," *Journal of Planning Education and Research* 19 (1999): 53–66. Emily Talen, "Traditional Urbanism Meets Residential Affluence: An Analysis of the Variability of Suburban Preference," *Journal of the American Planning Association* 67 (2): 199–216. Shlay, "Taking Apart the American Dream: The Influence of Income and Family Composition on Residential Evaluations."
6. Brower, *Good Neighborhoods: A Study of In-Town and Suburban Residential Environments.* Robert J. Sampson, "Linking the Micro- and Macrolevel Dimensions of Community Social Organization," *Social Forces* 70 (1) (1991): 43–64. A.B. Shlay and D.A. DiGregorio, "Same City, Different Worlds: Examining Gender and Work-Based Differences in Perceptions of Neighborhood Desirability," *Urban Affairs Quarterly* 21 (1985): 66–86. Daphne Spain, "The Effect of Changing Household Composition on Neighborhood Satisfaction," *Urban Affairs Quarterly* 23 (4) (1988b): 581–600. A.B. Shlay, "Castles in the Sky: Measuring Housing and Neighborhood Ideology," *Environment and Behavior* 17 (5) (1985): 593–626. Shlay, "Taking Apart the American Dream: The Influence of Income and Family Composition on Residential Evaluations."
7. American Lives, *New Urbanism Report* (San Francisco, CA.ENS-news, 2003). Poll Finds Environmental Concern on the Decline. April 17, 2003. Available at www.gallup.com/poll/releases/pr030417.asp. Michael Cabanatuan, "Bart Searches for Answers after Retrofit

Defeat," *San Francisco Chronicle*, November 14, 2002. Report based on a survey by the Public Policy Institute of California conducted by Mark Baldassare.

8. David Myers, *The American Paradox: Spiritual Hunger in an Age of Plenty* (New Haven: Yale University Press, 2000).
9. Daphne Spain, "An Examination of Residential Preferences in the Suburban Era," *Sociological Focus* 21 (1) (1988a): 1–8. David J. O'Brien and Stephen Ayidiya, "Neighborhood Community and Life Satisfaction," *Journal of the Community Development Society* 22 (1) (1991): 21–37.
10. Brower, *Good Neighborhoods: A Study of In-Town and Suburban Residential Environments*. Brian J.L. Berry, "Islands of Renewal in Seas of Decay," in Paul E. Peterson (ed.), *The New Urban Reality* (Washington, D.C.: Brookings Institution, 1985), 69–96. Ivonne Audirac, "Stated Preference for Pedestrian Proximity: An Assessment of New Urbanist Sense of Community," *Journal of Planning Education and Research* 19 (1999): 53–66. Donald N. Rothblatt and Daniel J. Garr, "Suburbia: An International Perspective on Levels of Satisfaction with the Physical Environment," *Journal of Planning Education and Research* 5 (2) (1986): 94–106. W.M. Michelson, *From Sun to Sun* (Totowa, NJ: Rowman and Allanheld, 1985). S. Fava, "Residential Preferences in the Suburban Era: A New Look?" *Sociological Focus* 18 (1985): 109–117.
11. Talen, "Traditional Urbanism Meets Residential Affluence: An Analysis of the Variability of Suburban Preference.".
12. Christopher Alexander, *The Timeless Way of Building* (New York: Oxford University Press, 1979). Jane Jacobs, *The Death and Life of Great American Cities* (New York: Random House, 1961). Lewis, Mumford, *The Urban Prospect* (New York: Harcourt, Brace & World, 1968). Kevin Lynch, *A Theory of Good Urban Form* (Cambridge, MA: MIT Press, 1984). William H. Whyte, *City: Rediscovering the Center* (New York: Doubleday, 1988). Jonathan Barnett, *The Fractured Metropolis* (Boulder, CO: Westview Press, 1995). Congress for the New Urbanism, *Charter of the New Urbanism* (New York: McGraw-Hill, 2000). Andres Duany and Elizabeth Plater-Zyberk, *Towns and Town-Making Principles* (New York: Rizzoli, 1991). Philip Langdon, *A Better Place to Live: Reshaping the American Suburb* (Amherst, MA: University of Massachusetts Press, 1994).
13. Jack L. Nasar, *The Evaluative Image of the City* (Thousand Oaks, CA: Sage, 1997). Nelessen, *Visions for a New American Dream*. George Peterson, "Measuring Visual Preferences of Residential Neighborhoods," *Ekistics* 23 (1967): 169–173. Stephen Kaplan and John S. Wendt, "Preference and the Visual Environment: Complexity and Some Alternatives," Environmental Design: Research and Practice. EDRA 3/AR 8, Los Angeles, 1972. Brower, *Good Neighborhoods: A Study of In-Town and Suburban Residential Environments*.
14. Kevin Lynch, *The Image of the City* (Cambridge, MA: MIT Press, 1960). Amos Rapoport, *The Meaning of the Built Environment: A Non-Verbal Communication Approach* (Tucson: University of Arizona Press, 1982). Allan B. Jacobs, *Looking at Cities*. (Cambridge, MA: Harvard University Press, 1985). John A. Jakle, *The Visual Elements of Landscape* (Amherst, MA: University of Massachusetts Press, 1987). Lawrence J. Vale and Sam Bass Warner, Jr., *Imaging the City: Continuing Struggles and New Directions* (New Brunswick, NJ: Center for Urban Policy Research, 2001).
15. For copyright protection issues, figures 8.2 and 8.4 are similar, but not identical to, the original images used in the survey.
16. Ralph B. Taylor, "Neighborhood Responses to Disorder and Local Attachments: The Systemic Model of Attachment, Social Disorganization, and Neighborhood Use Value," *Sociological Forum* 11 (1) (1996): 41–74. Richard E. Adams, "A Tale of Two Cities: Community Sentiments and Community Evaluation in Indianapolis and Pittsburgh," *Sociological Focus* 25 (3) (1992): 217–240. Oddvar Skjaeveland, Tommy Garling, and John Gunnar Maeland, "A Multidimensional Measure of Neighboring," *American Journal of Community Psychology* 24 (3)

(1996): 413–435. D.W. McMillan and D.M. Chavis, "Sense of Community: A Definition and Theory," *Journal of Community Psychology* 14 (1986): 6–23. Talen, "Traditional Urbanism Meets Residential Affluence: An Analysis of the Variability of Suburban Preference."

17. Julian Rotter, "Locus of Control Scale," *Psychology Today* (June 1971): 42. L. Festinger, *A Theory of Cognitive Dissonance* (Stanford, CA: Stanford University Press, 1957).
18. Talen, "Traditional Urbanism Meets Residential Affluence: An Analysis of the Variability of Suburban Preference."
19. Ibid.
20. The Knowledge Networks panel (see Methodology) is representative of the U.S. population as a whole in terms of race and ethnicity; however, panel response rates from minorities on this particular survey were low.
21. Gerda R. Wekerle, "From Refuge to Service Center: Neighborhoods That Support Women," *Sociological Focus* 18 (2) (1985): 79–95. Daphne Spain, "The Effect of Changing Household Composition on Neighborhood Satisfaction," *Urban Affairs Quarterly* 23 (4) (1988b): 581–600.
22. Information on this company is available at http://www.knowledgenetworks.com.
23. The association is negative because of the way the ranking of depending variables—the higher the number, the less favorable the respondent.
24. Favorable is used here as an all-encompassing term. The five semantic scales define favorable in different terms, for example, "beautiful," "exciting," and "safe."
25. Not shown in the table because the associated probability was 0.045.
26. Talen, "Traditional Urbanism Meets Residential Affluence: An Analysis of the Variability of Suburban Preference.".
27. All stores are included in the statistics given: Discount Stores, Supercenters, SAM's Clubs, and Neighborhood Markets. Information was obtained from www.walmart.com.
28. L. Festinger, *A Theory of Cognitive Dissonance* (Stanford, CA: Stanford University Press, 1957).
29. James Howard Kunstler, *Home from Nowhere* (New York: Simon & Schuster, 1996).

CHAPTER NINE

Fundamentalism and Antiurbanism: The Frontier Myth, the Christian Nation, and the Heartland

EDUARDO MENDIETA

Introduction: Two Nations

A few months after the terrorist attacks on the United States, David Brooks published an essay in *The Atlantic Monthly* entitled "One Nation, Slightly Divisible." In this essay Brooks offers an analysis of the electoral map that emerged after the 2000 election: a big block of red that stretched across the so-called Heartland surrounded on both coastal and northern edges by intense blue. The red heartland here refers to the Midwestern states, parts of South and Southwest, areas that are rural, agrarian, racially and ethnically homogeneous and religious. The blue coastal and northern areas are more urban, racially and ethnically mixed, and that tend to be more pluralistic and secular. Brooks aims to provide a colorful and polemical characterization of the type of person that inhabits each respective "nation." The heartland is recognizable because it is a place without "Starbucks, no Pottery Barn, no Borders or Barnes & Nobles. No blue *New York Times* delivery bags dot the driveways on Sunday mornings. In this place people don't complain that Woody Allen isn't as funny as he used to be, because they never thought he was funny. In this place you can go to a year's worth of dinner parties without hearing anyone quote an aperçu he first heard on *Charlie Rose.* The people here don't buy those litter rear-window stickers when they go to a summer vacation spot so that they can drive around with 'MV' decals the rest of the year; for the most part they don't even go to Martha's Vineyard."[1] Brooks then proceeds to expand on these juxtapositions. What is done with motors in red states is done without them in blue states. While churches dominate the former, Thai restaurants are prevalent in the other. While in red states Wal-marts are gigantic and have parking lots the size of state parts, in blue states the stores are "small but the markups are big." When characterizing the people who live in each region, Brooks puts it this way:

> We in the coastal metro blue areas read more books and attend more plays than the people in the Red heartland. We're more sophisticated and cosmopolitan—just ask us about our alumni trips to China or Provence, or our interest in Buddhism. But don't ask us, please, what life in Red America is like. We don't know. We don't know who Tim LaHaye and Jerry B. Jenkins are, even though the novels they have co-written have sold about 40 million copies over the past few years. We don't know what James Dobson says on his radio program, which is listened by millions. We don't know about Reba or Travis. We don't know what happens in Mega-Churches on Wednesday evenings, and some of us couldn't tell you the difference between a fundamentalist and an evangelical, let alone describe what it means to be a Pentecostal.[2]

Hyperbole aside, and rhetorical excesses suspended, there is something important in Brook's caricatures of red and blue states. What makes Brooks' schematic interesting is certainly not its accuracy, or even diagnosis of the sources of the apparent disuniting and fracturing of the nation. What makes Brook's piece interesting is the way in which it taps into a long-established mythology about the United States. What Brook is doing by invidiously comparing red with blue states is appealing to a longstanding myth of the U.S. national imaginary. At the center of this national imaginary is the myth of the frontier, a receding line where civilization and wilderness meet and humans are tested, denuded of the disabling sophistications of urban culture, and stripped down to the bare elements of true character. It is clear that Brooks is offering us a new version of the myth of the frontier, but now in terms of the red and blue ethos that inform the rural versus the urban divide that has emerged in the past half a century. At the same time, Brooks is right to underscore that while the red states are more pious, devoted, religious, the blue states are more secular, cosmopolitan, intellectual, and pluralistic. As Brooks put its, while people in the red states tend to live in "small towns, or small cities far away from the coasts," people in blue states live in and around "big cities on the coasts" (p. 53). With this comparison, Brooks has perhaps unwittingly given expression to two aspects of the myth of the frontier and how they have mapped what Leo Marx called the "moral geography" of the American national imaginary:[3] the conflict between the countryside and the city, and the concomitant correlation between religiosity and antiurbanism.

The tensions, enmity, and distrust between red and blue states can be expressed succinctly in terms of the antiurban prejudice of religious fundamentalism in the United States. The colonization of the United States, the expansion West, the conquest of the frontier, were guided, energized, and led by religious missionaries and a missionary ethos, but also by religious revivals. U.S. religious vitality was proportional to the expansion into the receding frontier and the conquest of the frontier wilderness. The religious project of evangelization and the national project of conquest of the Wild West converge in the national myth of the pastoral simplicity, honesty, humility, and naturalness of the frontiersmen. The true American became the frontier's ploughman, with his simple, primitive Christianity. The myth of the frontier has always entailed a moral geography that has mapped both practically and metaphorically the topography of the United States in terms of a national character that assigns to the countryside and the town true Christianity and the real "American," while relegating to the city secularism, vice, and those who are not real Americans. Religious fundamentalism is the other side of an antiurban ethos that today has metastasized into the red and blue national fracture that Brooks has caricatured. In order to understand better this new expression of this national mythology, we will go back to the earliest and most eloquent articulator of the myth of the frontier, Fredrick Jackson Turner. Then, we will explore how this myth has influenced a consistent antiurban attitude of the most important intellectual and cultural figures that have shaped the U.S. imaginary. In a fourth section, we will briefly look at Arthur M. Schlesinger's critique of Turner's thesis and his attempt to restore the place of cities in the evolution of the United States. Finally, in the fifth section, we will conclude with a look at the anti-intellectualism of the religious fundamentalism in the United States and its most recent incarnations in the mega-churches of the Midwest.

Richard Slotkin wrote in his magisterial *The Fatal Environment: The Myth of the Frontier in the Age of Industrialization 1800–1890*, "Myth is history successfully disguised as archetype."[4] The myth of the frontier disguised the history of the expropriation and genocide of the Native Americans, as it disguised the history of the rise of the United States as a mega-urban world. Myths disguise actual history, but they also conceal from us those forces that condition and guide our social existence. The myth of the frontier disguised and concealed from us the way in which the United States has been a society formed, fashioned, guided by the rise and development of its great cities. So long as we remain bewitched by the structuring influence of the myth of the frontier, U.S. antiurbanism will continue to synergize with religious anti-intellectualism, nativism, and political conservatism.

The Frontier and Americanization: Frederick Jackson Turner

In July 12, 1893, at a meeting of the American Historical Association in Chicago, Frederick Jackson Turner delivered a paper that has been without question one of the most formative and influential pieces of scholarship. The paper was entitled "The Significance of the Frontier in American History,"[5] and in it Turner proclaims a new interpretation of U.S. history. The essay was in fact a manifesto that rejected all U.S. historiography as it had been written up to the end of the nineteenth century. Turner explicitly takes a stand against the two reigning historiographical paradigms of the time. Neither slavery, nor the acculturation of European institutions, can give an appropriate analysis of the forces that have shaped the American character. Against both paradigms Turner juxtaposes the frontier, its continuing receding, and its continued conquest:

> American social development has been continually beginning over again on the frontier. This perennial rebirth, this fluidity of American life, this expansion westward with its new opportunities, its continuous touch with the simplicity of primitive society, furnishes the forces dominating American character. The true point of view in the history of this nation is not the Atlantic coast, it is the Great West. Even the slavery struggle, which is made so exclusive an object of attention by writers like Professor von Holst, occupies its important place in American history because of its relation to westward expansion.

Already in this paragraph Turner touches on a series of themes that will continue to structure how he sees the frontier as providing the *true point of view* of the history of the United States. For Turner, the frontier has secured the fluidity of U.S. institutions, making the nation ever young and fluid, always in the process of renewing and revitalizing itself. We also have the theme that the frontier is a place where U.S. society is stripped down to the bare essential, brought into contact with simplicity that has moral connotations. In addition, implicit in Turner's topographical shift from the Coast and the South to the West, is also a shift away from what has been called the national sin of slavery, toward another biblical image, Adam. For Turner, the slavery question was only an incident, and not a formative force. Indeed, implicit in Turner's frontiers men, who in his continuous touch with the "simplicity of primitive society," is reverted to a natural morality, an Arcadian, pastoral ethos, that was properly named by R.W.B. Lewis as the "American Adam."[6] The frontiers men, and the ploughmen of the westward expansion, are innocent of the genocide and expropriation of the Native Americans that had to be killed and pushed in order to keep pushing the frontier line. For Turner is quite explicit: "The most significant thing about the American frontier is that it lies at the hither edge of free land."[7] Although the line of the frontier has gone through successive relocations, its edge is always made up of free land, which paradoxically Turner acknowledges was "won by a series of Indian wars."[8] The frontier thus appears in Turner's topography as free land, as land without history, a *tabula rasa* in which the new American will imprint his moral character and on which his true character will be tested. In this way, then, Turner's topography projects a "moral geography" that maps, to use the term as Amy DeRogatis urges us to in her impressive *Moral Geography*,[9] onto the national landscape regions of moral exculpation and testing. The frontier thus is charged with theological connotations, for it is there where the American Adam goes to redeem his soul, to be cleansed of urban artificiality and strip down to the elemental virtues of primitive society.

Just as Turner rejects the formative and determinative role of slavery in the formation of the American character, he also rejects the thesis that the United States is essentially an appropriation of European institutions and character. Turner writes, and this merits lengthy citation:

> In the settlement of America we have to observe how European life entered the continent, and how America modified and developed

> that life and reacted on Europe. Our early history is the study of European germs developing in an American environment. Too exclusive attention has been paid by institutional students to the Germanic origins, to little to the American factors. *The frontier is the line most rapid and effective Americanization.* The wilderness masters the colonists. It finds him a European in dress, industries, tools, modes of travel, and thought. It takes him from the railroad car and puts him in the birch canoe. It strips off the garments of civilization and arrays him in the hunting shirt and the moccasins. It puts him in the log cabin of the Cherokee and Iroquois and runs an Indian palisade around him. Before long he has gone to planting Indian corn and plowing with a sharp stick; he shouts the war cry and takes the scalp in orthodox Indian fashion. In short, at the frontier the environment is at first too strong for the man... Little by little he transforms the wilderness, but the outcome is not the old Europe, not simply the development of Germanic germs, any more than the first phenomenon was a case of reversion to the Germanic mark. *The fact is, that here is a new product that is American.*[10]

In the frontier, the European immigrant is transformed into a true American. The frontier then became for Turner the liminal space of both death and birth, a kenotic space in which the old self is emptied and eviscerated so that a new self can be born. It is a liminal space with two edges. On one side, the European approaches and is transformed into a new man, a true American. On the other side, the wilderness is approached by what Turner calls the "disintegrating force of civilization" (p. 13) leading to the disintegration of the Indian way of life. From civilization the Indians acquired guns and alcohol, which led them to resist the European with greater force and determination in a pyrrhic struggle that only augured their own demise. While the frontier spells the demise of the Indian, and the freeing of the land on which the new nation will be established and reestablished, the frontier becomes the mythological space in which the frontiersman becomes the American Adam: sinless, virtuous, simple, unfettered by the vices, artifice, and accoutrements of duplicitous urbanity.

In January of 1903, in an essay published in the *Atlantic Monthly*, Turner reaffirms his analysis of the determining role of the frontier in U.S. history. In this essay, it is no longer the frontier as such, but the West in general. There is an important metonymic shift from the frontier to the West, in which now the West itself has become the kenotic space of American mythogenesis. In this essay, however, Turner does broaden his analysis of the social situation of the United States. He does speak about three other factors along with the West, as being determining for the future of U.S. democracy. Turner lists the concentration of wealth that can lay down the foundations of a new stage of development in the United States. Most of the concentration of this wealth has been matched by the concentration of labor with the rise of the new industries. Most of this labor, Turner points out, has been made up of immigrants. The next factor that Turner profiles is the "expansion of the United States politically and commercially into lands beyond the seas."[11] The final factor that Turner mentions is the national parties tend to divide on "issues that involve the question of Socialism."[12] Yet, at the head of Turner's list of factors that the United States faces the most significant is the closing of the frontier and the exhaustion of free land to be colonized. Thus, in 1903, for Turner the challenge was how was the United States to both preserve and revitalize its democracy when its primary source of vitality had exhausted. The challenge is articulated in this way: "This, at least, is clear: American democracy is fundamentally the outcome of the experiences of the American people in dealing with the West. Western democracy through the whole of its earlier period tended to the production of a society in which the most distinctive fact was the freedom of the individual to rise under conditions of mobility, and whose ambition was the liberty and well-being of the masses... The problem of the United States is not to create democracy, but to conserve democratic institutions and ideals."[13] But since the "*free lands* that made the American pioneer have gone,"[14] one must ask how is American democracy to conserve and revitalize itself. Turner's answer is "Let us see to it that the ideals of the pioneer in his log cabin shall enlarge into the spiritual life of a democracy where civic power shall dominate and utilize individual achievement for the common good."[15] Yet, one cannot fail to note the contradiction in this promise. For the ideals of the pioneer in his log cabin, as Turner ceaselessly repeated, are to a large extent antisocial, or rather, desocializing. The very ideals of the frontiersmen, the pioneer, are those that stand in opposition to the civic ideals of city dwellers, with their urbanism and civility. But just as there is a metonymic shift from the frontier to the West, there is a parallel shift from the frontiersmen as a person in the frontier, to the frontiersmen as an archetype that should now orient and enlarge the "spiritual life" of American democracy. The frontiersman becomes the quintessential American archetype, the germinal and formative national archetype.

Without doubt, Turner's description of the frontier and the frontiersman sketched a popular, patriotic, and self-congratulatory image that is most appealing and difficult to challenge or question. At the heart of this image are the admirable virtues of individualism, inventiveness, reverence of freedom and self-reliance, perseverance, humility and loyalty, and piety. Turner's celebration of the frontier rang, as Wilbur R. Jacobs put it in his foreword to the 1986 edition of Turner's essays on *The Frontier in American History*, "a historical freedom bell" that has not stopped ringing. Yet, as strong and appealing as Turner's thesis has been, its hazards and contradictions have remained inchoate and unacknowledged. For in Turner's thesis, the frontier turns into a myth that has religious consequences but that also makes it almost impossible to discuss the future of American democracy in a postfrontier society. Turner's celebration of the moral superiority of the frontiersman is directly related to his pastoralism or agricultural primitivism. Thus, a moral pastoralism, in the tradition of Jefferson, converges with a religious sanctioned derision of the urban. Henry Nash Smith eloquently analyzed the negative consequences of Turner's entwining democracy with moral pastoralism, when he noted that "What then was to become of democracy [after the free land of the frontier had disappeared]? The difficulty was the greater because in associating democracy with free land he had inevitably linked it also with the idea of nature as a source of spiritual values. All the overtones of his concept of democracy were therefore tinged with cultural primitivism, and tended to clash with the idea of civilization... Since democracy for him was related to the idea of nature and seemed to have no logical relation to civilization, the conclusion implied by his system was that post-frontier American society contained no force tending toward democracy."[16]

The Urban Republic: Arthur M. Schlesinger, Sr.

In 1933 Arthur M. Schlesinger, the senior, published *The Rise of the City: 1878–1898*, as volume 10 of a series he coedited with Dixon Ryan Fox.[17] The time span in the subtitle is misleading, for Schlesinger set out to offer more than a survey of the evolution of U.S. cities in such a short span. Rather, the book cover is a panoply of social issues that to this day make the book a pioneer in the field of U.S. Urban history and theory.[18] The book offered a social history of cities with an unblinking eye on labor history. For Schlesinger, the challenges of the rapidly industrializing and urbanism nation had to be analyzed in terms of the new social dynamics that were unleashed in cities. Yet, Schlesinger's book, notwithstanding the undertheorization of the city, also reflected on the emergence of a unique form of social interaction with attendant forms of psychic life and intellectual attitude. In an inchoate way, Schlesinger approached a phenomenology of urban existence that sketches the contours of an ethos and mental attitude that was open to differences, that craved intellectual growth, and that reveled in the new and foreign. For Schlesinger, the city nurtured a "spirit of impersonal social responsibility."[19] The rapidly growing city, in fact, entailed the emergence of social challenges for which the rural mentality and ethos were unprepared and insufficient. The rise of the city, in fact, coincided with a "golden age of invention," in the United States, which made it the most dynamic and fast-growing republic on the planet. One of the consequences of this new urban culture, with its mental attitudes, dynamism and inventiveness was that Schlesinger saw how the country was dividing into two cultures. While one was relegated to the rural areas was "static, individualistic, and agricultural," the other was nascent in the cities and was "dynamic, collectivistic, and urban."[20] Furthermore, one of these cultures was fated to lose the struggle for the allegiance of the nation. As cities were growing, more people were migrating to the cities than were going to the rural areas of the nation. For Schlesinger, the imminent triumph of the city was evident in the shift in the attitude of U.S. citizens toward the question of the Federal Government and the role of the stage in general. Citizens were no longer asking "how much authority the general government possessed," but where rather asking about "the alternative uses to which expanded power should be put"[21] (p. 416).

In 1940 Schlesinger returned to the role of the city in U.S. history with an essay entitled "The City in American Civilization," which he republished in an expanded and edited version in his 1949 collection of essays *Paths to the Present*.[22] This essay is particularly useful for our purposes because it is framed in terms of a critique of Frederick Jackson Turner's frontier thesis. Schlesinger opens the essay with a quote from Turner's famous essay, and then proceeds to note that in 1925 Turner himself had come to acknowledge the role of cities in the development of the nation, when Turner himself wrote that "an urban reinterpretation of our history" was need.[23] The essay, in contrast to his book *The Rise of the City*, offers in condensed form the structures for both an intellectual and social history of the United States. For instance, Schlesinger notes the role of cities in the emergence and coordination of the revolutionary spirit that lead to U.S. independence. Indeed, Britain's colonial policy after 1763 was a direct attack on the "roots of urban prosperity," leading to the unrest and dissatisfaction that would lead to independence. Here, Schlesinger focuses on Benjamin Franklin, as both the paragon of the new U.S. mentality and the urban ethos that will guide the emergent nation into a new stage in social evolution. Benjamin Franklin is a thoroughly urban intellect, which nourished not just on London and Paris, but also Boston and Philadelphia. While Franklin's name stands for an essential pragmatism, experimentalism, and empiricism, he is also an urban social scientist. Schlesinger writes: "Few elements of American culture but are indebted to his fostering care: printing, publishing, journalism, belles-lettres, education, the postal service, theoretical and applied science" (p. 214). Franklin was not just a statesman, scientist, philosopher, printer, but also and perhaps above all a civic leader with a thoroughly urban outlook that resulted in public lending libraries, local hospital and the first American Philosophical Society, to be hosted in Philadelphia, the first capital of the new Republic.

In evident contrast with Turner, Schlesinger reads the westward expansion of the United States, not just as the conquest of the frontier, but also as the process of the urbanization of the frontier. For Schlesinger, who had a more economic inflected understanding of U.S. history, the frontier was about the acquisition of resources that then could be injected into both national and international market, by means of expanding communications networks. The frontier would have been meaningless without the advancing army of roads and train-truck constructors. Curiously, Schlesinger as the United States expanded westward, the West was assimilated and connected more with the Eastern seaboard than with the south. This is an important dynamic that is going to lay down the factors that would lead to the Civil War. As the West, Midwest, and the East became more integrated into networks of fluvial and terrestrial commercial networks, the South remained caught in its own regional dynamic of slave labor and plantation culture. This makes Schlesinger's analysis of U.S. history not only more economically perspicacious, but also more nuanced. For Turner, the south and slavery are episodic and epiphenomenal, almost insignificant in the matrix of U.S. history. For Schlesinger, U.S. rise to world dominance cannot be understood without a proper analysis of the economic and urban dimensions of the frontier and its integration into the commercial networks of the Northeast: Boston, New York, Baltimore, as they linked to Chicago, Pittsburgh, and Cincinnati.

What Schlesinger notes repeatedly throughout his seminal essay is the contrast between the rates of growth of the rural versus the urban areas. Thus, while in 1776 1 in 25 colonists lived in areas with eight thousand or more inhabitants, the urban areas provided for the kind of concentration that allowed for a revolution. Between 1800 and 1860, the number of urban dwellers grew by a factor of 20, while rural folk grew only by a factor of 4. "By 1810 one out of every twenty Americans lived in communities of eight thousand or more, by 1840 one out of every twelve, by 1860 nearly one in every six" (p. 216). After the Civil War the rate of urban growth was double that of rural areas. Between 1860 and 1900, one in every six U.S. citizens now inhabited communities of eight thousand or more. By 1880 one out of four and by 1900 one out of three citizens were living in urban centers of eight thousand or over (pp. 223–224). One of the consequences of the Civil War and the abolition of slavery is that now the Northern and Midwestern pattern of urbanization would become national. It could be said that Lyndon B. Johnson's "war on poverty" program was aimed at urbanizing the still rural South. Yet, perhaps the most important number that Schlesinger provides us with is this one: "In the century from 1790 to 1890 the total population had grown 16th fold while the urban segment grew 139-fold" (p. 225). To which he adds, in direct reference to Turner's first sentence of his celebrated essay "The significance of the frontier in American history": "Hence the celebrated announcement of the Superintendent of the Census in 1890 that a frontier line no longer existed can hardly be said to have marked the close of 'the first period of American history.' Rather it was a tardy admission that the second period was already under way" (p. 225).

If Turner had sought to provide us with a new paradigm for understanding U.S. history, Schlesinger showed, without refuting directly Turner, that the frontier was not and could not have been the key to understanding U.S. development. Instead, Schlesinger showed that the United States had been from its colonial origins always an urban democracy, one that always integrated the rural areas into its forms of life, but also its dynamism. Interestingly, Schlesinger makes the point of arguing that it was the urban centers that revitalized democracy and the not the other way around as Turner had argued in many of his essays. Schlesinger concludes his essay with a call to a social task that given the continued antidemocratic and anti-intellectual spirit of the rural areas of the United States, remains as valid as it was in 1949 when he enunciated it: "... the American city leaped into being with breathtaking speed. At first a servant to an agricultural order, then a jealous contestant, then an oppressor, it now gives evidence of becoming a comrade and co-operator in a new national synthesis. Its economic function has been hardly more important than its cultural mission or its transforming influence upon rural conceptions of democracy. The city, no less than the frontier, has been a major factor in American civilization. Without an appreciation of the role of both the story is only half told" (p. 233).

The Intellectual versus the City: Morton and Lucia White

Richard Slotkin writes in the first volume of his magisterial trilogy on the myth of the frontier that "Myth-making... is simultaneously a psychological and a social activity. The myth is articulated by individual artists and has its effect on the mind of each individual participant, but its function is to reconcile and unite these individualities to a collective identity."[24] Slotkin writes these sentences in a section entitled "mythogenesis" in which he is articulating the hermeneutical key through which to understand the rise, development, and eventual stabilization and endurance of a myth that has continued to influence the national imaginary to our days. Evidently, individuals must not only be socialized into this imaginary, but they themselves have to at a sociopsychological level feel invested in this myth. The myth is both individual and social. It is a device through which we make sense of our social existence, but it is also the means through which individual biographies are made sense of. Slotkin's *Regeneration through Violence* is a masterful analysis of the ways in which the individual and social imaginaries weave and interweave in the myth of the frontier, with all of its attendant by-products. Yet, it is another work that has with equal mastery archived and excavated the mythogenesis of the frontier, but now with a particular focus on the enmity between the countryside and the city: Morton and Lucia White's *The Intellectual versus the City: From Thomas Jefferson to Frank Lloyd Wright.*[25] This comprehensive study is to be read in conjunction not just with Slotkin's trilogy but also with Richard Hofstadter's still indispensable *Anti-Intellectualism in American Life,*[26] to which we will turn later on. The Whites cover most of the intellectual history of the United States, focusing on some of the key figures in that history. The central animating preoccupation is explicitly articulate in the following way: "We have no persistent or pervasive tradition of romantic attachment to the city in our literature or in our philosophy, nothing like the Greek attachment to the *polis* or the French writer's affection for Paris" (pp. 1–2). In fact, there is no tradition of attachment to the city, because there is a strong and overwhelming tradition of deriding, critiquing, despising the city. It is this tradition of the derision of the city that the Whites trace in their study and given its breath and scope, I can only focus on some highlights, which I selected in light of our immediate purposes in this chapter.

The Intellectual versus the City opens properly with a chapter that explores the views on the city by three foundational figures in U.S. intellectual history: Franklin, Crèvecoeur, and Jefferson. It is without question, however, that the founding father of U.S. pastoral antiurbanism, and what has been also called "agricultural primitivism," is Thomas Jefferson. Jefferson inaugurated this tradition when he wrote the following words, which are some of his most quoted ones:

> In Europe the lands are cultivated, or locked against the cultivator. Manufacture must therefore be resorted to of necessity not choice, to support the surplus of their people. But we have an immensity of land courting the industry of the husbandman. It is best then that all our citizens should be employed in its improvement, or that one half should be called off from that to exercise manufactures and handicrafts for the other? *Those who labour in the earth are the chosen people of God, if ever he had a chosen people, whose breasts he has made his peculiar deposit for substantial and genuine virtue.* It is the focus in which he keeps alive that sacred fire, which otherwise might escape from the face of the earth. Corruption of morals in the mass of cultivators is a phaenomenon of which no age or nation has furnished an example.[27] (Italics added.)

Jefferson writes this in query XIX, in which he is addressing the issue of manufactures in his *Notes on the State of Virginia*, which he wrote in 1781 and expanded in 1782. Jefferson is here arguing that in contrast to Europe, the colonies have ample land to make them independent from the news of large-scale manufacture. In contrast to Europe, America does not need manufactures and the commerce that they entail. But most importantly, Jefferson attaches a moral dimension to the task of the husbandman, the farmer: they are the truly chosen of good, if God has a chosen people. For it is in tending to the soil that true virtues are schooled. It is this self-sufficiency of the American farmer that guarantees the preservation of their virtue. Were they to become dependent and subservient to the city, they would lose the ground of their pious virtuosity. In addition, it is the independence and self-sufficiency that secures the permanence of democratic self-governance. But how Jefferson concludes the remarks on query XIX is as telling:

> The mobs of great cities add just so much to the support of pure government, as sores do to the strength of the human body. It is the manners and spirit of a people which preserves a republic in vigour. A degeneracy in these is a canker which soon eats to the heart of its laws and constitution.[28]

Here the city becomes a synecdoche for the decease that eats at a democracy and an embodiment of the opposite of what the "free land" of the frontier means for the United States, both its independence and self-sufficiency that makes it be able to dispense from the necessity of large manufacture. In a memorable way, Jefferson has linked industry and urbanization with despotism and venality, while approvingly linking land, farming, and homestead independence, with moral virtue and democratic vitality. In fact, in a letter from September 23, 1800, addressed to Benjamin Rush, Jefferson will make his antiurbanism even more explicit and extreme: "I view great cities as pestilential to the morals, the health and the liberties of man. True, they nourish some of the elegant arts, but the useful ones can thrive elsewhere, and less perfection in the others, with more health, virtue & freedom, would be my choice."[29] At best, cities are barely tolerable. For the most part they are cankers, pestilential, sources of the ills of society and democracy. As the Whites note, "In Jefferson's eyes the republic and the city joined hands only in a marriage of convenience, with no thoughts of love at all."[30]

If in Jefferson we encountered a political-moral derision of the city, when he turns to another giant of U.S. letters, Emerson, we now encounter a transcendental, metaphysical, epistemological rejection of the city. In Emerson, the "distaste for the city" is directly linked to his theory of knowledge. Emerson had distinguished between reason and understanding, following Immanuel Kant. According to Emerson Understanding, "toils all the time, compares, contrives, adds, argues; near-sighted but strong-sighted, dwelling in the present, the expedient, the customary."[31] One could almost complete this list with other equally invidious adjectives: pecuniary, commercial, penny-pinching, vulgar, pedestrian, puny, and crass. In contrast, Emerson defines reason in the following way: "never reason [calculates and argues], never proves; it simply perceives; it is vision" (p. 25). As the Whites put it, "Reason was the soaring faculty of the philosopher and the poet, while Understanding was that of the ordinary, lumbering scientist" (p. 25). But, what is more important and telling is that for Emerson Understanding is associate with the city, with Reason is characteristic of the countryside. Emerson writes: "the city delights the Understanding. It is made up of finites: short, sharp, mathematical lines, all calculable. It is full of varieties, of successions, of contrivances. The country, on the contrary, offers an unbroken horizon, the monotony of the endless road, of vast uniform plains, of distant mountains, the melancholy of uniform and infinite vegetation; the objects on the road are few and worthless, the eye is invited ever to the horizon and the clouds. It is the school of Reason."[32] If understanding can only grasp the beauty of the city, a beauty that is instrumental and pecuniary, reason can rise to the sublime, that which exalts the mind and is the highest faculty of the soul. For Emerson, cities not only corrupt the mind, but also the character, for they make us talkative and entertaining, and thus artificial and duplicitous. The countryside, in contrast, maintains our moral transparency and uncorrupted behavior. The city is the perdition of the soul, the countryside its proper sanctuary.

Reading the Whites' monograph makes for chastening and sobering reading. Eloquently and sweepingly they track the characteristics of a deeply ingrained antiurbanism in the fabric of the American intellect. Only one figure in their study seems to have risen about the animus against one of the most important and evident aspects of U.S. society, its cities, and that was William James. For James, the conflict between the "tender-footed Bostonians" and the "Rocky Mountain toughs" is to be mediated by pragmatism, a philosophy that is thoroughly urban and procivitas. James' way of articulating in his Columbia University lectures on *Pragmatism* the conflict between the two mental types is particularly notable, not just because he seems to be challenging Emerson's epistemological topology, but also because it is challenging a deeply ingrained juxtaposition, that was poignantly expressed by Turner with his frontier thesis and by Jefferson with his encomium to the husbandman, and which was installed as an epistemic and ontological principle by Emerson. But what is also notable is that his defense of the pragmatic synthesis of the tender-minded and the tough-minded is articulated in terms of the juxtaposition between Leibniz's urban theodicy—which is expressed in the sentence "The evil will appear as almost nothing in comparison with the good, if we consider the real magnitude of the city of God"[33]—and the "valiant anarchist writer" Morrison I. Swift's reports from the suffering of the city attacking Leibnizian optimism. The pivot of the juxtaposition is the suffering of human in the city. At stake is what I called Leibniz's urban theodicy. For James, pragmatism promises a type of thinking that can negotiate between the Pollyannaish Leibnizian acceptance of suffering and incredulous and cynical rejection of any attempt to make sense of the social reality of human suffering. Pragmatism is the true philosophy of the urban social reality in which we all find ourselves. The Whites summarized James' proto-prourbanism thus: "A livable city on earth, one is therefore tempted to say, is the social manifestation of James' pragmatism, and that is why he is one of the first great American philosophers to associate himself with the effort to accept what is good and to root out what is bad in the life of the American city." (p. 145)

The Moral Geography of God's Country: Fundamentalism, Antiurbanism, and Anti-Intellectualism

Thus far I have been reading closely some key U.S. thinkers in order to articulate the links that exist between the "spatial and moral order,"[34] to use the felicitous expression by Amy DeRogatis. These links can be nicely expressed with the title of "moral geography." I have shown how Turner's frontier's thesis projects a moral geography onto the imperial mapping of the United States. At the heart of that moral geography is also a theology, in such a way that we can also speak of a geographical theology that entails both a theodicy and a salvation history. The redemption of the nation is to be sought in the trial and tribulation of the frontier line. The frontier turns into a quasi-sacred space, one in which the moral transgressor gets to be cleansed and saved. Here a brief reference to Sophocles' *Oedipus at Colonus* would be instructive.[35] After blinding himself and being exiled from his city for his parricide that is also a regicide, and incestuous relation with his mother, Oedipus comes upon the "grove of Eumenides," a sacred space that is the entrance to the netherworld. It is on this spot that Oedipus receives sanctuary and ultimately is turned into quasi-divine superhuman hero.[36] Turner's frontier line is not unlike Sophocles's grove of Eumenides: sinful, fallen, lacking, and wanting human, victims of their vice, foibles, but also arrogance, encounter refuge and ultimate redemption there. And this is how the frontier has operated in American history, structuring its imaginary and mythology. That, notwithstanding the very obvious fact that the frontier was hardly an independent factor in U.S. history and what that came to have any significance precisely because of its urban rearguard. This is what Arthur M. Schlesinger showed so eloquently and indisputably in the essay from 1949 that we glossed above. Yet, it is with the work of the Whites that we caught a more stark look at the entwinement between the moral geography of the frontier line with its pastoralism and agricultural primitivism and a celebration of U.S. religious identity. It was Jefferson that articulated in unequivocal terms the link between pastoralism, national identity, and antiurbanism. In the moral geography of Jefferson's pastoralism we have the synergy of religious fervor with antiurbanism that culminates in the twentieth century with the anti-immigrant, Know-Nothing Party, and the anti-urbanism of the religious revivals of the early part of century. It is this very same moral and theological geography that informs the fundamentalist and evangelical movements of the past half a century.

Fundamentalism is a term that dates back to between 1910 and 1915, when *The Fundamentals; or Testimony to the Truth*, the collective title for a series of booklets that were published with the support of Lyman and Milton Stewart. These pamphlets were distributed throughout the United States to protestant leaders. In them, well-known protestant theologians set out to defend the "fundamental" principles of U.S. religion not just against the onslaught of modernism, but also of the liberal and corrosive influence of the city immigrant.[37] In the early part of the century, fundamentalists became well known because of their rejection of Darwin, brought to national prominence in the Scopes trial, but also all forms of the introduction of science in curriculums of school. Since the early part of the twentieth century, fundamentalists have maintained a ceaseless struggle against modernism, secular and urban culture. In more recent times, fundamentalists have spawned even more extreme forms of antimodernism with even more militant stances against the vices of the city and urban culture. Richard Hofstadter captured succinctly the dynamics that have directed the development of Protestantism and Evangelicals in twentieth-century U.S. history: "The older, rural and small town America, now fully embattled against the encroachments of modern life, made its most determined stand against cosmopolitanism, Romanism, and the skepticism and moral experimentalism of the intelligentsia. In the Ku Klux Klan movement, the rigid defense of Prohibition, the Scopes evolution trial, and the campaign against Al Smith in 1928, the older America tried vainly to reassert its authority; but its only victory was the defeat of Smith, and even that was tarnished by his success in reshaping the Democratic Party as an urban and cosmopolitan force, a success that laid the groundwork for subsequent Democratic victories."[38] In short, and as the premier historian of fundamentalism put it, its mental outlook emerged from "the clash of two worlds, the rural and the urban."[39]

According to a poll conducted by political scientist John Green in 2004, about 12.6 percent of U.S. citizens defined themselves as "traditional evangelicals." This poll, however, is based on a response to a definition offered by the pollster. In contrast, according to the measures indicated by a poll lead by George Barna, almost 40 percent of Americans are born-again Christians, of which only 7 percent are true evangelicals. Now, born-again Christians are those who claim a personal relationship with Jesus Christ and also accept him as their sole savior. In contrast, evangelicals are those who additionally believe that "their faith is very important in their life today; believe they have a persona responsibility to share their religious beliefs about Christ with non-Christians; believe that Satan exists; believe that the eternal salvation is possible only through grace, not works; believe Jesus Christ lived a sinless life on earth; and describe God as the all-knowing, all-powerful, perfect deity who created the universe and still rules today."[40]

It is this active and powerful minority that conservatives and republicans have called the "moral majority," and they have been the ones to decide elections over the past half a decade in the United States, perhaps only until the election in 2008 of Barack Hussein Obama.[41] But at the core of their appeal and endurance are those themes that echo back to Jefferson, Emerson, and Turner: the true American in the heartland, toiling the land, conquering the frontier, an internal spiritual frontier, where the true American dwell and is forget in the crucible of the testing wilderness. The difference between those claim they live in the so-called pro-American states of the country and those from the early part of the century, who did live in rural areas, is that these born-again Christians, evangelicals, and fundamentalists have become exurban, that is exiles from the cities of Babel, Sodom, and Gomorrah. In an irony that should not escape this new soldier of god, they seek refuge in the suburbs of urbanized rural America, in the mega-churches of the heartland, where they gather under steel and glass structures to be connected via satellite with like-minded believers. Hofstadter characterized the fundamentalist mentality as one that is essentially Manichean, one that "looks upon the world as an arena for conflict between absolute good and absolute evil, and accordingly it scorns compromises (who would compromise with Satan?) and can tolerate no ambiguities."[42] But as I have argued this Manichean worldview is the other side of a profoundly antiurban outlook that sees all evil as gathered in and emanating from the city.[43]

In a powerful essay on one of the mega-churches that is carrying on Christ's crusade against the modern sinful work, Jeff Sharlet writes: "As contemporary fundamentalism has become an exurban movement, it has reframed the question of theodicy—If God is good, then why does He allow suffering?—as a matter of geography. Some places are simply more blessed than others. Cities equal more fallen souls equal more demons equal more temptation, which, of course, leads to more fallen souls. The threats that suffuse urban centers have forced Christian conservatives to flee—to Cobb Country, Georgia, to Colorado Springs. Hounded by the sins they see rampant in the cities (homosexuality, atheistic school teaching, ungodly imagery), they imagine themselves to be outcasts in their own land. They are the 'persecuted church.' "[44] In the ultimate analysis, as Schlesinger and countless social critics have pointed out, the pastoralism and agricultural primitivism that is so attractive and definitional to U.S. citizens is parasitic on the city. The city itself has urbanized the rural areas making them in suburbs. "Myth does not argue its ideology, it exemplifies it" argued Slotkin.[45] Now that we have a century of argument against the evident fallacy and falsity of the pseudo-pastoralism and pseudo-ruralism of the United States, we can perhaps begin to finally put away the myth of the frontier with its moral and theological geographies in which true Christian and Americans are embattled against the city and its racially mixed cosmopolitan citizens. Already at the turn of the century, when the frontier had long been conquered, the future of the United States was the future of its cities. Today that is truer than ever. The future viability and vitality of U.S. democracy resides in the right to the city that the antiurbanism of United States's most evoked and dear myth has made impossible.

Notes

1. David Books, "One Nation, Slightly Divisible," *Atlantic Monthly*, December 2001, 53.
2. Ibid., 53.
3. Leo Marx, *The Machine in the Garden: Technology and the Pastoral Ideal in America* (New York: Oxford University Press, 1964), 72.
4. Richard Slotkin, *The Fatal Environment: The Myth of the Frontier in the Age of Industrialization 1800–1890* (Middletown, CT: Wesleyan University Press, 1986), 20.
5. Frederick Jackson Turner, *The Frontier in American History* (Tucson, Arizona: University of Arizona Press, 1986 [1920]), 1–38.

6. R.W.B. Lewis, *The American Adam: Innocence, Tragedy and Tradition in the Nineteenth Century* (Chicago and London: Phoenix Books, 1955).
7. Turner, "The Significance of the Frontier," 3.
8. Ibid., 9.
9. Amy DeRogatis, *Moral Geography: Maps, Missionaries, and the American Frontier* (New York: Columbia University Press, 2002).
10. Turner, "The Significance of the Frontier," 4. Emphasis Added.
11. Frederick Jackson Turner, "Contributions of the West to American Democracy," in Frederick Jackson Turner, *The Frontier in American History*, 245–246.
12. Ibid., 246.
13. Ibid., 266.
14. Ibid., 245.
15. Ibid., 268.
16. Henry Nash Smith, *Virgin Land: The American West as Symbol and Myth* (New York: Vintage Books, 1950), 304.
17. Arthur Meier Schlesinger, *The Rise of the City: 1878–1898. A History of American Life*, Vol. 10 (New York: Macmillan, 1933).
18. See Terrence J. McDonald, "Theory and Practice in the 'New' History: Rereading Arthur Meier Schlesinger's *The Rise of the City, 1878–1898*," *Reviews in American History* 20 (3) (September, 1992): 432–445.
19. Schlesinger quoted in McDonald, "Theory and Practice in the 'New' History," 439.
20. Schlesinger, *The Rise of the City: 1878–1898*, 50.
21. Ibid., 416.
22. Arthur M. Schlesinger, *Paths to the Present* (New York: Macmillan, 1949), 210–233.
23. Schlesinger, "The City in American Civilization., 210.
24. Richard Slotkin, *Regeneration through Violence: The Myth of the American Frontier, 1600–1860* (Middletown, CT: Wesleyan University Press, 1973), 8.
25. Morton Gabriel White and Lucia White, *The Intellectual versus the City: From Thomas Jefferson to Frank Lloyd Wright* (Cambridge, MA: Harvard University Press and MIT Press, 1962).
26. Richard Hofstadter, *Anti-Intellectualism in American Life* (New York: Vintage Books, 1963).
27. Thomas Jefferson, *Writings* (New York: Library of America, 1984), 290. Italics added for emphasis.
28. Ibid., 291.
29. Ibid., 1081.
30. Morton and White, *The Intellectual versus the City*, 19.
31. Emerson quoted in Ibid., 25.
32. Ibid., 25.
33. Leibniz quote by William James in "Pragmatism" in *Writings 1902–1910* (New York: Library of America, 1987), 496.
34. DeRogatis, *Moral Geography*, 2.
35. See Sophocles, *The Three Theban Plays: Antigone, Oedipus the King, Oedipus at Colonus*, Robert Fagles (trans.) (New York: Penguin Books, 1982).
36. See Darice Birge, "The Grove of the Eumenides: Refuge and Hero Shrine in Oedipus at Colonus," *Classical Journal* 80 (1) (October–November 1984): 11–17.
37. See Randall Balmer and Lauren E. Winner, *Protestantism in America* (New York: Columbia University Press, 2002), 73–87. See also Mark A. Noll, *A History of Christianity in the United States and Canada* (Grand Rapids, MI: William B. Eerdmans, 1992), 381–386.
38. Hofstadter, *Anti-Intellectualism in American Life*, 123.
39. George M. Marsden, *Fundamentalism and American Culture: The Shaping of Twentieth-Century Evangelicalism: 1870–1925* (New York and Oxford: Oxford University Press, 1980), 185.
40. George Barna quoted in Michelle Goldberg, *Kingdom Coming: The Rise of Christian Nationalism* (New York and London: W.W. Norton, 2006), 9.

41. The "perhaps" is to indicate a hesitation that the dynamic that Brooks identified in the article I quoted at the outset may still be at work in U.S. elections. While Obama received 364 electoral votes, McCain received only 162. While Obama received 52.7 percent of the popular vote, McCain received 46 percent. But what is most telling are the ways in which those votes were spread out. Cities went overwhelmingly for Obama, while rural areas went over for McCain. Locales with over half a million inhabitants went 70 percent over to Obama, while only 28 over to McCain. As Marjorie Connelly put it in the an article in the times analysis the 2008 election, "Mr. Obama won majorities in the Northeast, Midwest and West. Mr. McCain won in the South, which republicans have won since 2000." Marjorie Connelly, "Dissecting the Changing Electorate," *New York Times*, Sunday, November 9, 2008, Week in Review, 5. See also Mike Davis, "Obama at Manassas" *New Left Review*, 56 (March-April 2009): 5–40, for a detailed analysis of the urban, exurban, and suburban patterns that lead to Obama's election.
42. Hofstadter, *Anti-Intellectualism in American Life*, 135.
43. See Jon C. Teaford, *Post-Suburbia: Government and Politics in the Edge Cities* (Baltimore: Johns Hopkins University Press, 1997).
44. Jeff Sharlet, "Soldier's of Christ: Inside America's Most Powerful Megachurch," *Harper's Magazine* 310 (1860) (May 2005): 41–54, quote at 50.
45. Slotkin, *The Fatal Environment*, 19.

CHAPTER TEN

*Against Safety, Against Security: Reinvigorating Urban Life**

DON MITCHELL

Antiurbanism takes many forms, economic as well as social, political as well as cultural. One way that antiurbanism—defined perhaps as an opposition to the *publicness* of the city—has become manifest in the spaces of the city itself is through an overweening concern with safety and security. One ideal of urbanism is precisely that it is *there*, in the city, that chance, serendipity, the unexpected, has the best chance to flourish, and out of chancy, serendipitous, and unexpected encounters new social, cultural, and political formations might arise. Rarely achieved in practice, this ideal of urbanism is nonetheless critical to contemporary theories of citizenship—as something to be struggled toward.[1] Control over city space, practices of safety and security, that is, has never been stronger to the production of urban space; but in recent history (and not just since 9/11) certain practices of security (separating out "strangers," mechanical and human surveillance, defensive design), have become, in many ways, defining. Discourses of safety have become dominant in discussions of "good" urban space. There is nothing wrong with being concerned with, and seeking to design in relation to, security, especially bodily security, or the safety of urban space users. But, I want to suggest in what follows, that when security and safety are defining, a certain antiurbanism rooted in fear—what I call "social agoraphobia"—comes to be the primary structuring force of urban life. I further suggest that this antiurbanism succeeds to the degree that it is profitable for some, often at the expense of others. For that reason, I think it is an obligation of progressive urban scholars to begin articulating theories that are against safety, and against security, if we want to promote something other than the antiurban city.

The fear that the securitization of public space induces and responds to is indeed a form of agoraphobia. "Agoraphobia" is the clinical name for the experience of overwhelming fear in public spaces. It is a serious mental disorder that affects considerably less than 1 percent of the population in the United States (and even fewer in Europe) (WHO 1992).[2] Agoraphobia is marked by an inordinate fear of being in crowds or in spaces that seem to be under the control of others. Sufferers often become trapped in their own homes, afraid to venture outside, even to go to the supermarket. It is a debilitating illness that can lead to a complete withdrawal from social life: fear to be *in* public space can come to define all that one does and doesn't do, and therefore all that one is.

Very, very few of us suffer from agoraphobia, at least in this sense. Even so, agoraphobia is coming to define our lives.[3] For us, this agoraphobia is not so much a fear to be *in* public space as an induced fear *of* public space. Upon this second kind of agoraphobia—the fear *of* public space—I will argue, much contemporary capitalism now develops. Widespread agoraphobia of the first sense—the sense of a debilitating fear to be *in* public space—would be a disaster for capitalism. Malls would empty out; hopping streets would be deserted; even our private lives would become more closed to commerce (and to the state) as we stopped answering the phone, refused to step outside to collect our mail, and even disconnected from the Internet for fear of who might be on line with us. By contrast, a widespread fear *of* public space is enormously productive for capitalism—and for the state formations that safeguard it.

We are taught fear *of* public space in many ways. News media sensationalize assaults and murders that take place in public (while minimizing the true extent of domestic violence). They exalt the values of gated communities and sports utility vehicles that insulate their owners from the surrounding crowds. In schools we learn that anything public—including schools!—is suspect at best, and more likely dysfunctional, while everything private is efficient, clean, and to be wondered at. We learn that private property is the foundation of all freedom, and even, by the time we go to university and sit through our first economics and politics classes, we learn that freedom is not just impossible, but in fact inconceivable, without private property. We learn to make important distinctions. We learn that "public" is the same as out of control: public spaces are the realm of criminal violence, homeless people, drugs, anarchy, terrorists; public hospitals are where one goes to find long lines and waiting lists; public schools "fail our children" (as American politicians like to put it); and public goods are, by definition, simply inefficient. The private, on the other hand, is the very definition of good: publicly accessible private properties are where profit is made, comfort and security provided; private property, because always efficiently allocated, works *for* us all; private wealth benefits us all.

Social agoraphobia is crucial in this formulation. If clinical agoraphobia keeps people locked away indoors, social agoraphobia delivers us into the waiting arms of merchants in safe and secure malls, developers of securely gated neighborhoods, and newly redeveloped urban spaces like Times Square so carefully watched over by its army of private security guards and privately operated CCTV cameras.[4] Urban design, in other words, more and more serves both to *induce* fear (by so thoroughly separating everyday life from the totality of the social world of which it is part) and to *allay* that fear by providing spaces of sociability that feel urban at the same time they feel controlled, safe, surveilled, and almost entirely unthreatening.

Consider in this regard downtown San Diego, California. San Diego is a city where being outside, being in public, is—or at least should be—immensely pleasurable. The sun shines; it is warm year round; and even massive suburban growth and 1960s-style urban redevelopment failed to fully destroy the old, walkable downtown.[5] Horton Plaza is the old heart of downtown. It was once an open square, deeded by city developer Alonzo Horton in 1894 to create a central gathering place, something like a cross between a Latin American plaza and a New England town common. He stipulated in his deed to the city that it remain forever open and forever public. It was a ceremonial square, where politicians gave speeches on national holidays, and city grandees showed themselves off. It was also where strikers gathered and rallied against the bosses of the new city, where preachers sought to convince passersby of the wrath of a vengeful God—or the beneficence of a loving one—and where transient workers rested between jobs, trips to the saloon, and registering at the nearby employment offices.

By the 1960s, transient workers were gone, replaced by active and discharged service men from the nearby military bases, and especially by the elderly poor who lived in surrounding Single Room Occupancy (SRO) hotels, and used the tree-shaded benches of the plaza as an outdoor sitting room. Like the rest of the downtown (outside the skyscraper core) with its small residence hotels, cheap bars, and burlesque shows, Horton Plaza was a lively, if decidedly seedy, place. But it was also an increasingly class-specific one, as the bourgeoisie and the wealthy moved out of downtown to the new suburbs, La Jolla for the truly privileged, La Mesa for the merely middle class. Shopping followed the money and moved out too, to new malls in Mission Valley or up in La Jolla, and the old propertied elite of downtown realized that while they might be living quite comfortably in their mansion with a view over the beach, their investments in downtown property were in trouble. The seediness of Horton Plaza and its denizens, the large number of cheap hotels and apartments catering to the indigent, and so much property simply not being put to any conceivable "highest and best use," became a sore in the side of the city's elite. From about 1965, therefore, it's easy to track a steadily more insistent drum beat of complaint in the local press: downtown is unsafe; it's dirty; it's filled with the poor, with racial minorities, with illicit activities. It is violent. It didn't matter that downtown crime was no higher, and in some cases much lower, than other city areas. Vilification of downtown became something of a local sport.[6]

This history of San Diego, in other words, is exactly the same as so many other cities: the out-of-control New York of the 1970s, Glasgow in the 1980s, Vancouver in the 1990s. And like all these other cities, such vilification served an important purpose: once demonized, the city could be saved. In San Diego a newly revitalized redevelopment agency, controlled by large property owners, arranged to condemn much of the property around Horton Plaza and hand it over to a large suburban mall developer, the Hahn Corporation (builder of the wildly successful University Town Center in La Jolla that did so much to empty out downtown). Hahn built the festival marketplace-style Horton Plaza Shopping Center on the land handed over to him. As part of the deal, the City of San Diego rebuilt Horton Plaza Park, moving the bus stop that used to be next to it three blocks down the street (so that those waiting for the bus would no longer hang out in the park), fencing off the open lawn, removing the benches from in front of the new mall's main department store, and eliminating the public toilets that used to be below the park. The city even gave the name—Horton Plaza—to the Hahn Company, allowing it to trademark it, make it its own property. Nothing else could use the name Horton Plaza, so the plaza became Horton Park. (And when I talk to local San Diegans these days and just say "Horton Plaza" they *always* think I am talking about the shopping center.) The new downtown mall opened to much fanfare in 1984. Predictions were that it would save downtown.[7] In many ways these predictions were right (if we don't worry too much for now about what "save" might mean), but it took a while.

The poor and elderly, for example, failed to get the message and insisted on continuing to use the park. As did military service members. So the city arranged for the USO—an organization dedicated to providing welfare and recreation for military people—to move out of downtown. It stepped up its demolition of Single Room Occupancy Hotels. It got the Rescue Mission, providing services to the homeless, to move to what was then called Center City East, and in 1989, the City once again agreed to the redevelopment of the park. This time they used the Hahn corporation's own money (there was no pretence of public funding) to remove the lawn and *all* the benches. The lawn was replaced with prickly plants, the benches were simply not replaced. After this redesign, it simply became impossible to sit comfortably anywhere in the plaza. The park became a conduit to the Mall (where there was plenty of seating, at least at the restaurants, cafes, and fast food joints, and at least if you spent money in these places). After this redevelopment, there was no reason to stop in the park, and every reason now to hurry toward the open doors of the shops.

And so that's what everyone did—except of course the poor and homeless who were not welcome on the private property of the Horton Plaza Shopping Center. Private security guards patrolled the walkways and sitting areas of the mall, forcibly removing anyone they found to be "undesirable." The contrast between the clean, safe, open space of the private mall, and the now entirely unwelcoming space of the public park could not have been clearer. Indeed, it was made especially stark by the fact that no matter the destruction of all the Single Room Occupancy hotels, no matter the eviction of services for the homeless and elderly to other parts of town, homeless people, the indigent elderly, and other "down-and-outs" still often sat on the low curbs, or directly on the walkways, invoking a certain abject fear, or at the least disgust, in office workers as they tried to enter the mall at lunchtime. Once in the mall, once safely on private property, all was sparkle and splendor, cozy seats awaited, the abundance of the market beckoned, the prerogatives of a stylish city life were there for the taking—or the buying, at any rate.

In the meantime, the success of Horton Plaza Shopping Center, which rapidly began attracting tourists and suburban residents (who could drive to the mall and park in its garages without once ever setting foot on a city street or in Horton Park), had begun to have a spillover effect (just as hoped and planned) in the neighboring "Gaslamp

Quarter" (a gentrifying turn-of-the-century district), which rapidly developed into an upscale bar and restaurant district, with those Single Room Occupancy hotels that were not simply demolished magically transformed into upscale boutique hotels for wealthy tourists. Not incidentally, then, the gentrification of the Gaslamp forced hundreds or thousands of very poor, often elderly people, out of their homes and into the streets and the tattered shelter system of the city. So while a relatively small number of homeless or other poor people hanging out, however uncomfortably, in the redesigned Horton Plaza Park, had the rather positive effect of pushing more wealthy people even more rapidly into the space of the mall, the growing numbers of homeless in the city as a whole, coupled with the fact that the streets and sidewalks of the Gaslamp remained public property that the homeless had a presumptive right to be on, presented the city with a certain problem: homeless-induced fear *of* public space could easily become a hindrance to making money off the redeveloped landscapes of the city. If wealthy patrons (suburbanites, tourists, convention-goers) stayed away from the Gaslamp restaurants, or refused to cross Horton Plaza Park to get into the shopping center, the profit machine that San Diego was being turned into would grind to a halt.

Bound by the United States Constitution, no less (in many cases) than by their own knowledge of who was on the streets and why they were there, the public, city police felt constrained in their ability to move the poor out, to get them off the streets and out of the parks, to just to push them out of downtown. The public police could not, legally at least, bar the poor from the streets and parks open to the public. But from property owners and from the newly arriving upscale residential population of the 1980s and 1990s (drawn in by a veritable explosion of condominium construction) came a loud cry: "The homeless must go!" The answer, originally, was much the same as in the rest of the United States. In the midst of the Reagan and later neoliberal reforms, the new urbanists of San Diego, and their representatives in city government, saw the answer not in making room for the poor by providing affordable housing, and certainly not in striking a blow against poverty by seeking to reform an increasingly and glaringly unjust political economy, but in using *even more* privatization to push them out—to just make the poor and homeless disappear.

Horton Plaza Shopping Center was the model: on private property, poor and homeless people, or youth and sometimes even people of color, seeking to hang out in the comfortable confines of the mall, could simply be excluded (Staeheli and Mitchell 2008, pp. 59–70).[8]

They could be stopped at the entrance; curfews could be invented. If they snuck in, they could be forcibly removed. The homeless and loitering youth could be removed because private police—and for that matter the public police—could enforce the wishes of private property owners. But this is less easy on publicly owned streets. So in the late 1990s, the City Council voted to withdraw many public services (street cleaning, garbage collection, and some policing functions)—to just stop funding them—in the downtown area, and simultaneously to allow a private organization to collect a fee from business and property owners to provide these services. Beholden to their new fee paying clients, rather than to the Constitution or voters, and freed of other claims on these funds, the newly instituted Property and Business Improvement District (P-BID) increased street cleaning services. They also implemented new private police-like patrols, whose main job, as the President of the P-BID told my colleague Lynn Staeheli and myself in an interview, was "to get in the face" of homeless people and make it clear that they did not belong downtown—to "roust" them, as he put it.[9] The P-BID named these new services "Clean and Safe"—and that was its goal, to make the streets clean and safe for particular classes of people by getting rid of others.

And Clean and Safe could get away with this precisely *because* it was private. As private people, uniformed Clean and Safe guards were understood, by law, to be merely engaged in free expression when they told the poor and homeless to get out. As long as they did no bodily harm to homeless or other street people, who after all still had a putative right to be on public property, the private Clean and Safe patrols could do wonders in making it clear that street people were not welcome. They could induce in *them* a very well-grounded fear *of* (remaining in) public space.

With very little debate, then, much policing in San Diego has been privatized, using the public's fear *of* public space as a pretext. So successful has this been than when presented an opportunity (and indeed the legal obligation) to create a new park downtown, the City of San Diego simply turned the task of regulation to a private concern: the San Diego Padres baseball team. Like many American cities, San Diego wanted to build a new baseball stadium downtown. And like many American cities, it wanted to create the surrounding district in the *image* of the old public city, a city that provided public space, street life, and the sort of vibrancy generally considered to be at the heart of urbanism. And yet, at least ostensibly, very few people in San Diego really *wanted* public space—the history of the destruction of Horton

Plaza Park and the great success of Horton Plaza Shopping Center's pseudo-public space made that abundantly clear. What they wanted was something that *seemed* like public space, but really wasn't. The solution for San Diego was to assure that the park built in conjunction with the baseball stadium—the park that was to be the truth of the vibrant, public city—never *was* public. Rather, the City granted to the baseball team the right to come up with a plan for when the park would open, under what conditions, and who would be allowed in and who would be kept out. Right from the beginning, then, this city property and ostensibly public space was not public at all and city property in name only: it was a tightly controlled simulacra of public space geared just as much as was Horton Plaza to the making of money. After all, the San Diego Padres Baseball team was granted the right to close the park altogether when baseball games were being played to assure that no one would be able to see the game unless they paid the price of admission to the stadium. Developing the park this way has proved to be widely popular with the people of San Diego, who feel that by being privately controlled, the parks (as well as the stadium, the shopping center, and the bars and restaurant district of the Gaslamp) will necessarily be more safe.[10] One of the primary advocates for the homeless in San Diego, Father Joe Carroll of St. Vincent de Paul, the largest provider of services to the homeless in the city, told Lynn Staeheli and me that as a resident of the city, he wished all public spaces would be privatized and all the homeless kicked out.[11] And in the summer of 2007, a reporter for a weekly San Diego newspaper laughed out loud as she was interviewing me about San Diego's plans to create more downtown parks and I said that was all well and good, but they really ought to be *public* and not like the baseball park or what Horton Plaza had become. She thought I was being completely unrealistic (as sympathetic as she actually was to my argument).

I was being unrealistic because it is now just so utterly obvious, just so much common sense, in American cities, that the induced fear *of* public space is beneficial, something devoutly to be wished for than to fought against. People still want to be in public, but they only want to be in public if the public place they are in is private: if it is privately owned and privately policed. The benefits of such a state of affairs are obvious. Disney, for example, has become a large urban planner and developer, hired to remake whole streetscapes is Seattle and (most famously) to refashion New York's Times Square in an image of spectacular, now "family oriented," consumption.[12] It has also created Celebration, a fully planned, fully private town in Florida, where the streetscapes are supposed to evoke turn-of-the-twentieth century small American cities (where streets are easily walkable, full of people, and not dominated by the automobile), but where also, nonresidents can simply be barred from entering. All services are private, all laws and rules are written by Disney (and almost impossible to change by the residents) and all policing is geared toward maintaining not urban life but property values.[13]

Across the continent in Los Angeles, Universal Studios has built "City Walk," designed by the architect Jon Jerde, who also designed Horton Plaza Mall, and where for a fee one can wander a perfect urban scene complete with a few—but not too many—picturesque beggars (who are City Walk employees, dressed up in costume and required to follow a script in their interactions with visitors). Strollers in City Walk are offered a carefully controlled, and therefore guaranteed never to be disappointing, urban experience.[14] The long history of inducing fear *of* public space makes such developments seem not just logical, not just safe, but utterly desirable. It makes the shopping mall—with its every square meter privately owned and carefully managed to assure profit—seem like the highest achievement of urbanism. Horton Plaza Shopping Center is not just now the symbolic center of downtown San Diego, it *is* the center of downtown San Diego.

Fear *of* public space helps us realize that a life lived off the property of capital is a life not worth living. It induces a deep desire for space that is commodified, carefully designed, utterly predictable (in the way that a night at the Hard Rock Café, anywhere in the world, is utterly predictably and *therefore* fun—as a recent (2007) Hard Rock magazine advertising campaign says, "you know where to go." Where we want to go, apparently, is a *controlled* environment, because we know that whenever we are in a controlled environment, we are in a *good* environment.

And so this induced fear *of* public space, and this banishment of fear *from* controlled space that might just appear public, makes us want to have our fingers printed, our retinas scanned, our backpacks searched, and our credit limits preapproved. It doesn't just make us want these, really; what it does is make us *miss* this kind of surveillance when we are not subject to it. When we are not subject to the careful designs and even more careful surveillance of private capital (the same capital that awaits to sell us a bottle of water when we are thirsty even though there is a public tap right at hand; the same capital that entices us with a new sweater when we are a little depressed, even though our closets are already overly full), we start to feel vulnerable and exposed. We start to feel like we have no safe retreat. The streets around us, the palm-shaded Plazas we used to visit, all become places to be avoided. It might be better to just stay at home, with the lights dimmed, and maybe just read a book, or even just climb into bed and stay there and never come out—and, perhaps, even stop shopping.

Without the private police of the mall, that is, without the security guards hired by the Property-Based Business Improvement District—San Diego's "Clean and Safe" program—without the agents of "homeland security" probing into every aspect of our lives (and whisking to jail those who have the temerity to protest, and do research, against gentrification), without the geniuses of marketing always probing to find out what makes us comfortable (and thus willing to spend) and what makes us afraid (and therefore likely to bury our wallets even deeper in our pockets), we start to feel afraid, almost agoraphobic.

Thus by producing places that are public space's safe, clean, *better* simulacra, private property and its privatizing ally the neoliberal state, is there to assure, by making us afraid *of* public space, that we need not be afraid *in* the new, private, public space—that we never truly, and unprofitably, become agoraphobic.

* * *

But why should I, why should we, be against this? Why should we be against security, against safety? Surely one of the primary jobs of the state—of any state, and indeed of any society—is to assure the security and safety of its members. And surely, though there may be some few problems around the edges, a few pesky homeless people, for example, who have no place to be, no place to go, except the public spaces that no longer exist—the private provision of safety and security through the privatization of property is *efficient*, serving great goods to great numbers. One could, of course, make a valiant defense of those pesky homeless people, on classic liberal grounds having to do with the sanctity of individuals' human lives, or on more radical grounds having to do with the way that capitalism works and what it does to those who somehow cannot sell their labor power, and indeed I have tried in much of my work, to make these kinds of defenses.[15] But I think there is a broader, and necessary, if perhaps more abstract argument to be made against the forms of security and safety that capital and its client state is making for us in contemporary cities. This broader argument against safety and security is in fact a positive argument for a *right to the city* that begins by rejecting any induced fear *of* public space, even if, or perhaps because, some public spaces are, indeed, unsafe.

The idea of a *right* to the city is closely associated, of course, with Henri Lefebvre.[16] For him the right the city was a "cry and demand"—something vital, something necessary, something both under threat and in need of creation. The right to the city, for Lefebvre, was primarily the right to participate in the *making* of the city, the right to *shape* urban life, the right to what he called *centrality*—which is the right to be present and alive and at the heart of things. Lefebvre called the right to the city, as a right to *make* the city, the right to the *oeuvre*: the right to the city as a work or a project. The right to the city was thus always a "right-in-the-making," not a once-and-for-all established institution or attribute, not so much a liberty right that gives one a nearly unfettered ability to do as one pleases in some aspect of social life, but rather a form of power, a form of empowerment. The right to the city is, especially, a form of *collective* empowerment. As Lefebvre put it,

> Among these rights in the making features the *right to the city* (not the ancient city, but to urban life, to renewed centrality, to places of encounter and exchange, to life rhythms and time uses, enabling the full and complete *usage* of these moments and places, etc.).[17] (Italics in the original.)

Centrality, encounter, exchange, life rhythms: this is a language of use and belonging, of being in and part of a multifarious public. It is *not* merely the language of consumption and security. It is rooted, in fact, in a conception of the city as an ongoing project of difference in which *collective* use values predominate. It is therefore a conception of the city in which the possibility of *dis*order—the possibility for a certain *loss* of control, a certain *in*security—is always present and not necessarily entirely to be feared. This loss of control, this insecurity, is in itself productive of the right to the city.

Yet the right to the city is not necessarily an end in and of itself. According to Lefebvre, the end toward which the right to the city as *oeuvre* tends is, *la Fête*, which is a "celebration which consumes unproductively, without other advantage but pleasure and prestige and enormous riches in money and object." *La Fête*—the festival—for Lefebvre was a moment of Dionysian revel, in which everyday life was turned on its head (even as the festival was *part* of everyday life). As Andy Merrifield has put it, "During festivals, people dropped their veils and stopped performing, ignored authority and let rip."[18] One of the hallmarks of the twentieth century, according to the rock critic and social theorist Greil Marcus (1989), was how the festive spectacle became generalized through commodity production and was regularly enacted in a rock club, on TV, and now, in its particularly safe and secure mode, in the very spaces of San Diego's Horton Plaza Shopping Center, both diluting its force and making it more productive for exchange value than for use value. Nonetheless, and against this commodification of the spectacle, Lefebvre, to the end of his life, held out hope that the festival, as Merrifield put it, could be

> a special, potentially modern form of Marxist practice that could erupt on an urban street or in an alienated factory. The festival was a pure spontaneous moment, a popular "safety valve," a catharsis for everyday passions and dreams, something both liberating and antithetical: to papal infallibility and Stalinist dogma, to Hitlerism and free-market earnestness, to bourgeois cant and born-again bullshit.[19]

In other words, *la Fête*—and thus the city as *oeuvre*, and thus the right to the city—is *dangerous*: it is, indeed, *against* safety, and *against* security, at least as it is conceived in the contemporary city defined by the fear *of* public space. *La Fête* portends instead a world out of control, a world where the disempowered are empowered, and where safety and security take a back seat to joy and creativity—and to radically transformed geographical contexts.[20]

After all, as Lefebvre wrote elsewhere, the 1871 Paris Commune "was the biggest celebration of the century and modern times," and the only real "crack at revolutionary urbanism the modern world has faced"—which goes far in explaining the force of the reaction the Commune called up.[21] As David Harvey has shown so well, the reclamation of Paris by the forces of order required a reordering of space, a remaking of the imprint of power over and in public space.[22] It required, in a word, a reimposition of *fear* as a structuring force: a fear not only of the potential violence of the revolutionaries, and a fear not only of the actual and organized violence of the state, but especially a fear of *space itself*—a fear of that thing that revolutionary, or even everyday, people, possessed of a right to the city actually produce. A radical right to the city, in other words, is—and must be—the antithesis of fear, the antithesis of the social agoraphobia so carefully constructed in contemporary cities.

Agoraphobia—clinical or social—is a form of paranoia, and paranoia defines much contemporary life. Paranoia is the state of being "very distrustful or suspicious of others."[23] Paranoia is both a generalized condition and, as we have seen, a strategy—a strategy of control and a strategy for profit. It is also, in its landscape manifestations, utterly banal. What this paranoia *looks* like is San Diego's Horton Plaza—both the park and the shopping center. What it looks like is the so-called public park outside the San Diego baseball stadium. What it looks like are all the fee-based play zones that have sprung up to profit from parents to scared to let their children play outside. What it looks like is a CCTV camera on every corner, above every doorway, observing every space. What it looks like, perhaps, is Las Vegas. Las Vegas, in the last generation, has been "reinvented" as a "family resort."[24] But that is not the most interesting thing about it. Rather, the true attraction of Las Vegas is that every social act, and every *risk*, is scripted in advance. You know the odds, and you know they are not in your favor and that the house always wins, but that is better than not knowing what is around the next bend or about to happen in the park across the street.

In other words, the fear of homeless people that has driven both the redevelopment of Horton Plaza Park and the creation of the "Clean and Safe" program, or even just the amorphous fear of what Zygmunt Bauman aptly notes is often just apprehended as an unknowable "world out there," is in fact the fear of the unknown *within* a well-known context: the context of urban space.[25] The goal for planners then, in Vegas or in San Diego, is to change the context, to rework the space so that something other than a right to the city in the Lefebvrian sense can be made and maintained. The goal is to regiment space so that behaviors may be all the more carefully scripted, but to do so, as in Las Vegas, in such a way so that such new spaces are, indeed, just what we want, just what we desire. We know where to go.

Fear both *of* public space and *in* public space—the generalized social agoraphobia I have been talking about today—thus serves as a pretext for a larger re*ordering* of social interactions, and a hoped-for general surrender to authority,[26] a surrender that makes *la Fête*, except in a scripted bacchanal as in Las Vegas, not only unrealizable, but unthinkable.

This surrender to authority takes many forms. It is often expressed in polls through a simplistic assessment of how many, and what kinds of civil liberties people are willing to trade for a greater sense of security. Sometimes it is expressed as an argument against liberty itself. It is not uncommon here, certainly in the American case, that it is precisely the abandonment of rules, tradition, and "order," especially in the 1960s, that has led to the abandonment of all that is ethical and good. Too much liberty, this argument goes, has spawned nothing but mayhem (and people sleeping on the streets). For the conservative commentator Heather Macdonald, reflecting on the continuing crisis of homelessness in San Francisco, the problem is not that people cannot afford a place to live and shelter beds are woefully too few in number, rather, it is that San Francisco too ardently "pursu[ed] freedom" and what it got instead was "chaos." But, at the time she was writing, when there was a talk-tough mayor in office, she was heartened that San Francisco was "rediscovering that liberty consists not in overturning social rules, but in mutual adherence to them."[27] Alas, this did not solve homelessness either, even if it did manage to criminalize homeless people, but that is not the point.

The point is that the reinforcement of order itself comes to stand as a social good in and of itself. This love of order has strange effects. In 2003, the U.S. Supreme Court unanimously upheld the conviction of a young man in Richmond, Virginia named Kevin Hicks.[28] In this case, the Court declared that it was perfectly legal for a city council to transfer ownership of public streets to a semipublic agency—that is to take a step even further than San Diego has attempted with its Clean and Safe program, and hand the streets themselves right over to an unaccountable agency. In the Richmond case, when the city gave its streets to the local housing authority, it also required that authority to post no trespassing signs, and to arrest any person who did not have an (undefined) "legitimate social or business purpose." Once arrested for trespassing, and in fact even before then, a person could also be served with what was called a "trespass-barment notice" that banned her or him *forever* from the once-public streets.

The pretext for this rather drastic move was drug dealing, and also the war on terrorism if the oral arguments of the U.S. Solicitor General and the briefs of various state governments are to be believed—but the bid, of course, is to gain authoritarian control over the streets and to make them the functional equivalent of the private space of a mall—like Horton Plaza Shopping Center. In its decision the Court could not have been clearer, that the whole privatization and trespass-barment procedure was a ruse. Upholding the arrest and permanent barment of a young man who had otherwise done no harm (he was on his way to visit his girlfriend and children), the Court intoned:

> *Most importantly*, both the notice barment rule and the "legitimate business and social purpose rule" apply to *all* persons who enter

> the streets of Whitcomb Court [the housing estate whose streets had been privatized] .... The rules apply to strollers, loiterers, drug dealers, roller skaters, bird watchers, soccer players, and others not engaged in constitutionally protected activity... (Italics added.)

For the court, the only constitutionally protected activity is handing out leaflets and asking for signatures on a petition. What the Court is saying is that we have no a priori right to be on the streets unless (1) we have a legitimate business purpose; (2) we have a legitimate—that is state sanctioned—social purpose; or (3) we are engaged in legitimate—that is state sanctioned—political activities. We have no right to just hang out; we have no right to the city. *This* is what an over-weaning concern with safety and security gets us.

One of the most important aspects of this Supreme Court case is its endorsement of *trespass* laws in public space.[29] By endorsing the expansion of trespass rules into public property, the Court is creating a powerful tool for the maintenance of safety and security and the destruction of rights to the city. This tool is a tool of purification and pacification; it is a tool of *spatial* control in that it is designed to carefully vet and regulate *where* people are the better to control *what* they do.

This is not at all a new strategy, but it is one gaining in force. Keeping people out—out of nation-space as well as city space—is now a primary task of any government. And no wonder. For, by the 1990s, according to the great British satirist, we had all become aware of "Global Warning." "Everywhere," Brown says:

> There was a rise in Global Warning. Every day, there were new Global Warnings about killer viruses, killer waves, killer drugs, killer icebergs, killer vaccines, killer killers and other possible causes of imminent death. At first these Global Warnings were frightening, but after a while people came to enjoy them.[30]

Or at least some people did. If you were one of those imminent threats, not killer vaccines, but killer, or just jobless, immigrants, who might also be killer terrorists, or you were a homeless person who merely wanted a place to sit, even if you smelled like shit, then nation-, continent-, and city space became a fortress from which you were to be absolutely excluded. And on city streets, the tool of choice for securing public space against the likes of you is the enforcement of no trespassing laws. In the city, no trespass laws seem to be the last stay against the

Global Warnings Craig Brown sees as determining everyday life. They are certainly the last stay against the likes of Kevin Hicks, whose main crime you might be interested to know, was seeking to bring diapers to his girlfriend and baby. Or at least that was his official crime. His real crime, of course, was that he was young, and male, and black, and *therefore* a perceived threat to the safety and security of the streets.

Given this, engaging in the act of trespass, in the city of safety and security, takes on new meaning, new importance. Trespass is not only a "crime against property" (as if it is possible to commit a crime against a mere thing), and it is not only an act of disobedience (and such perceived disobedience to authority is certainly what drew the ire of the local judge when Kevin Hicks was first arrested). Rather, to trespass is now, precisely, to "make an unwarranted claim, intrude, encroach," not just upon a person's time or attention, their patience or their property (as the dictionary would have it), but especially upon the authority of the state and capital.[31] Hanging out in Horton Plaza Park or walking the streets of the Whitcomb Court housing project is truly an *unwarranted* claim. To be "warranted" means "permitted by law or authority, authorized, justified, sanctioned"[32] and there is almost nothing left that we can permissibly do in our everyday urban lives that is not, now, warranted. Without permission, without authorization, we are always intruders; we are always intruders against and on the safety and security of the spaces we seek to occupy. And—crucially—*so is everyone else.* Until authority can be established and shown, we are all now potential trespassers, each making a claim on another's attention, time, property, and sense well-being.

It is enough to make one agoraphobic. Paranoia threatens to reign supreme, and the only way to corral it is to reconstruct the world as one in which the no trespassing sign, and the desire for the safety and security it seems to represent, the *willing* surrender to a higher authority that it truly signals, is welcomed with open arms and deeply internalized. The only way to corral this paranoia, to reassure our desire for safety and security is to create a world in which trespass in foundational to governance. And, so, privatization of urban space—the creation of Clean and Safe and Horton Plaza Shopping Center *and* the handing of public parks over to private corporation—together with the privatization of urban life, or really the destruction of urban life, that this entails becomes something truly to be desired. Private means safe; private means secure. Public is nothing more, and nothing less, than anarchy.

As early as 1961, Henri Lefebvre warned of the "reprivatization of life"[33]—a warning echoed repeatedly by, for example, the urban sociologist Richard Sennett.[34] The answer for both Sennett and Lefebvre was to re*inhabit* the city *as* a city, and—*against* the overweening power of the no trespass sign, *against* safety, *against* security—to reassert those *urban* values of difference, strangeness, and danger. We must reinhabit not only the city, but city spaces, recognizing fear and danger not as something that can, or *even that should* be vanquished, but something that has to be *lived*. To learn to live in this fear, to assert the values of the urban despite it, and to fight instead for a world of productive difference where trespasser—that world of warrants and authorization—takes a backseat to inclusion and joy, where safety is subordinate to the frisson of urbanity, and where security plays second fiddle to the right to the city as an *oeuvre*, to do that is the challenge that faces us not only as radical urban scholars, but even more as people who live in the here and now and want to live somewhere better.

Notes

A different version of this chapter was printed in the proceedings of the 2007 Deutsche Geographentag, *Umgang mit Risiken: Katastrophen, Destabilisierung, Sicherheit* (*Coping with Risk: Catastrophe, Destabilization, Security*) (Bayreuth 2007).

1. See, for example, Iris Marian Young, *Justice and the Politics of Difference* (Princeton: Princeton University Press, 1990); Kurt Iveson, *Publics and the City* (Oxford: Blackwell, 2007); Sophie Watson, *City Publics* (London: Routledge, 2006).
2. World Health Organization, *ICD-10: Classification of Mental and Behavioural Disorders* (Geneva: World Health Organization, 1992).
3. The following discussion of agoraphobia, and the case study of San Diego, is a revision of an argument made in Don Mitchell, "Frogkt og Konstructions av det Offentlige Rom," *Adbusters Magasin* (Norway) 1 (2004): 24–27.
4. See Kristine Miller, *Designed on the Public: The Private Lives of New York's Public Spaces* (Minneapolis: University of Minnesota Press).
5. A good overview of the historical geography and political economy of San Diego can be found in Mike Davis and Jim Miller, *Under the Perfect Sun: The San Diego Tourists Never See* (New York: New Press, 2003); see also Lynn Staeheli and Don Mitchell, *The People's Property?: Power, Politics, and the Public* (New York: Routledge), Chapter Three.
6. Ibid., 49–51.
7. Ibid., 51–62.
8. Ibid., 59–70.
9. Ibid., 68.
10. J. Heller, "Squeeze Play on Park Space Rejected; Council Tells Padres to Rework Proposal," *San Diego Union-Tribune*, April 24.
11. Staeheli and Mitchell, *The People's Property?*, 48.
12. Miller, *Designs on the Public*; Stacy Warren, "Disneyfication of the Metropolis: Popular Resistance in Seattle," *Journal of Urban Affairs* 16 (1994): 89–107.

13. Andrew Ross, *The Celebration Chronicles: Life, Liberty and the Pursuit of Property Value in Disney's New Town* (New York: Ballentine, 1999).
14. A. Reevew and R. Simmonds, "'Public Realm' as Theatre: Bicester Village and Universal City Walk," *Urban Design International* 6 (3–4) (2001): 173–190.
15. Don Mitchell, *The Right to the City: Social Justice and the Fight for Public Space* (New York: Guilford, 2003).
16. Henri Lefebvre, "The Right to the City," in E. Kofman and E. Lebas (eds. and trans.), *Writings on Cities* (Oxford: Blackwell), 61–181.
17. Ibid.
18. Andrew Merrifield, *Henri Lefebvre: A Critical Introduction* (New York: Routledge, 2006), 14.
19. Ibid., 17–18.
20. Henri Lefebvre, *The Urban Revolution* (Minneapolis: University of Minnesota Press, 2003).
21. Quoted in Don Mitchell and Joaquin Villanueva, "The Right to the City," in R. Hutchison (ed.), *The Encyclopedia of Urban Studies* (Thousand Oaks: Sage, forthcoming).
22. David Harvey, *Paris: The Capital of Modernity* (New York: Routledge, 2003).
23. *Shorter Oxford English Dictionary*: Paranoia.
24. Hal Rothman, *Neon Metropolis: How Las Vegas Started the 21st Century* (New York: Routledge, 2003).
25. Zygmunt Bauman, *Liquid Fear* (Cambridge: Polity, 2006).
26. See Clayton Rosati, *The Image Factory: MTV, Geography, and the Industrial Production of Culture*, PhD Dissertation, Department of Geography, Syracuse University.
27. Quoted in Mitchell, *The Right to the City*, 181.
28. The following discussion of, and all quotations concerning, the Hicks case is derived from Don Mitchell, "Property Rights, the First Amendment, and Judicial Urbanism: The Strange Case of *Hicks v. Virginia*," *Urban Geography* 26 (2006): 565–586.
29. For an analysis of the rise of trespass in relation to governing homelessness, see Don Mitchell and Nik Heynen, "The Geography of Survival and the Right to the City: Speculations on Surveillance, Legal Innovation, and the Criminalization of Intervention," *Urban Geography* (Forthcoming).
30. Bauman, *Liquid Fear*, 5.
31. Definitions from *Shorter Oxford English Dictionary*: Trespass.
32. *Shorter Oxford English Dictionary*: Warrant.
33. See Merrifield, *Henri Lefebvre*.
34. See, for example, Richard Sennett, *Families Against the City* (New York: Vintage, 1974); *The Fall of Public Man* (New York: Norton, 1992); and *Flesh and Stone* (New York: W.W. Norton, 1996).

CONTRIBUTORS

Robert A. Beauregard is a Professor in the Graduate School of Architecture, Planning, and Preservation at Columbia University. He is the author of *Voices of Decline: The Postwar Fate of U.S. Cities* (New York: Routledge, 1993 and 2003) and *When America Became Suburban* (University of Minnesota Press, 2006) and was coeditor with Sophie Body-Gendrot of *The Urban Moment: Cosmopolitan Essays on the Late Twentieth-Century City* (New York: Sage, 1999). His work focuses on both planning theory and urban theory with particular attention to decline and redevelopment.

Angela M. Blake is Associate Professor in the Department of History at Ryerson University in Toronto, Canada, where she teaches courses in U.S. history, with a special focus on urban cultural history. She is the author of *How New York Became American, 1890–1924* (Johns Hopkins University Press, 2006). Blake is working on a second book entitled *Audible City* that investigates the political and cultural impact of changes in the post-1945 urban soundscapes of New York City and Los Angeles.

James A. Clapp is Emeritus Professor of Planning and Urban Studies at San Diego Sate University. He has taught at the University of California, the University of Paris and at Lingnan and City Universities in Hong Kong as a Fulbright Scholar in 2000. He has also lectured at several universities in Beijing and Shanghai and was recently made an "honorary professor" at Beijing City University. He has also won awards for his work in both print and broadcast journalism and has been an international tour guide. In addition to well over one hundred articles, scripts, and reviews he is author of nine books. His most recent book is his first work of full-length fiction, published under the name Sebastian Gerard, *For Goodness Sake, A Novel of the Afterlife of Suzie*

Wong. Email: urbmedia@mac.com. Web site: www.homepage@mac.com/yingloon.

Leo Marx is a Professor in the department of Program in Science, Technology, and Society at MIT. He received his received his B.A. (History and Literature, 1941) and his Ph.D. (History of American Civilization, 1950) from Harvard University. He taught at the University of Minnesota and Amherst College before coming to MIT in 1976. He has three times been a Fulbright Lecturer in Europe, twice a Guggenheim Fellow, and a Rockefeller Fellow. He is a fellow of the American Academy of Arts and Sciences, and has been president of the American Studies Association, and chair of the American Literature Section of the Modern Language Association. He is the author of *The Machine in the Garden: Technology and the Pastoral Ideal in America* (1964); *The Pilot and the Passenger: Essays on Literature, Technology, and Culture in America* (1988); editor (with Merritt Roe Smith), *Does Technology Drive History?: The Dilemma of Technological Determinism* (1994); and editor (with Bruce Mazlish), *Progress: Fact or Illusion?* (1996).

Elena Mancini is an independent scholar residing in New York City. She received her Ph.D. in German Studies from Rutgers University in 2007 with a dissertation entitled "Magnus Hirschfeld, The Third Sex and the Sexual Freedom Movement in Germany." She is currently coauthoring a book on women's workplace rights. She has taught at Rutgers University and Hunter College, CUNY.

Eduardo Mendieta is Professor of Philosophy at the State University of New York, Stony Brook, where he is also the director of the Center for Latin American and Caribbean Studies. He is the author of *The Adventures of Transcendental Philosophy* (Rowman & Littlefield, 2002) and *Global Fragments: Globalizations, Latinamericanisms, and Critical Theory* (SUNY Press, 2007). He is presently at work on another book entitled *Philosophy's War: Logos, Polemos, Topos*. His most recent book publications are a collection of interviews with Angela Y. Davis, entitled *Abolition Democracy: Beyond Empire, Torture and War* (Seven Stories Press, 2006), and an edited volume of interviews with Richard Rorty, *Take Care of Freedom, and Truth Will Take Care of Itself* (Stanford University Press, 2006). Two of his most recent article publications are "Prisons, Torture, Race: On Angela Y. Davis's Abolitionism" *Philosophy Today, Supplement* (2006): 77–82, and "War the School of Space: The Space of War and the War for Space," *Ethics, Place and Environment* 9 (2)

(June 2006): 207–229. Professor Mendieta was the executive editor of *Radical Philosophy Review* and is presently on the advisory board of the American Philosophical Association. He has coedited a book entitled *Biopolitics and Racism: Foucauldian Genealogies* (SUNY Press) that offers in translation some key texts in Foucault's development of the concept of biopolitics, as well as a collection of essays that apply this concept to U.S. social realities.

Don Mitchell is a Distinguished Professor of Geography in the Maxwell School at Syracuse University. After receiving his Ph.D. in Geography from Rutgers University in 1992, he taught at the University of Colorado before moving to Syracuse. He is the author of *The Lie of the Land: Migrant Workers and the California Landscape* (1996); *Cultural Geography: A Critical Introduction* (2000); and *The Right to the City: Social Justice and the Fight for Public Space* (2003) as well as numerous articles on the geography of homelessness, labor, urban public space, and contemporary theories of culture. His latest book, with Lynn Staeheli, called, *The People's Property? Power, Politics, and the Public* (2008) has just been published by Routledge. He is currently working on a new NSF-funded project called: *Bracero: Remaking the California Landscape, 1942–1964*. Mitchell is a recipient of a MacArthur Fellowship and has held a Fulbright Fellowship in the Institute for Sociologi og Samfunnsgeografi at the Universitetet i Oslo and has been a Visiting Scholar at the Annenberg School for Communication at the University of Pennsylvania. He is the founder and director of the People's Geography Project (www.peoplesgeographyproject.org) and a member of the Syracuse Hunger Project.

Roger A. Salerno is Professor of Sociology at Pace University and Chair of the Sociology/Anthropology Department. He served as a city planner in New York City and worked in East Harlem in the area of community development. His interests are in social theory and psychoanalysis. He latest book *is Sociology Noir: Studies at the University of Chicago in Loneliness, Marginality and Deviance* (McFarland, 2007).

Emily Talen is a Professor in the School of Geographical Sciences, with affiliations in the Schools of Sustainability and Planning, at Arizona State University. She holds a Ph.D. in urban geography from the University of California, Santa Barbara and a master's degree in city and regional planning from Ohio State University, and is a member of the American Institute of Certified Planners. Her recently released book, *The Design of Diversity* (Architectural Press, 2008), explores the urban design requirements of socially diverse neighborhoods in Chicago.

Michael J. Thompson is Assistant Professor of Political Science and serves on the faculty of Urban Studies at William Paterson University. His most recent books include *The Politics of Inequality: A Political History of the Idea of Economic Inequality in America* (Columbia University Press, 2007) and *Confronting the New Conservatism: The Rise of the Right in America* (New York University Press, 2007).

INDEX